Is Science
Necessary
?

Is Science Necessary

Necessary?

ESSAYS ON SCIENCE AND SCIENTISTS

?

MAX F. PERUTZ

BARRIE & JENKINS
LONDON

First published in Great Britain in 1989 by
Barrie & Jenkins Ltd
289 Westbourne Grove, London W11 2QA

British Library Cataloguing in Publication Data
Perutz, M.F. (Max F.), 1914–
Is science necessary?: essays on science and scientists.
1. Society. Role of science.
I. Title.
303.4'83

ISBN 0-7126-2123-7

DESIGNED BY EARL TIDWELL

Printed in the United States of America

For permissions and acknowledgments see pages xi to xiii,
which constitute an extension of this page.

To my grandchildren, Timothy and Marion,
and their future.

It is science alone that can solve the problems of hunger and poverty, of insanitation and illiteracy, of superstition and deadening custom and tradition, of vast resources running to waste, of a rich country inhabited by starving people. . . . Who indeed could afford to ignore science today? At every turn we have to seek its aid. . . . The future belongs to science and those who make friends with science.[1]

JAWAHARLAL NEHRU (1889–1964)
First Prime Minister of the Free India

CONTENTS

ACKNOWLEDGMENTS

I should like to thank the following for permission to reproduce figures and tables: Her Majesty's Stationery Office (figures 2, 7, and 28); Dr. M. W. Service and Academic Press (figure 8); Dr. N. R. Lardy and *Scientific American* (figures 4 and 5); Food and Agriculture Organization of the United Nations (figures 3 and 6, table 1); Dr. A. M. Anderson and *Nature* (figure 13); Bundesverband der Pharmazeutischen Industrie (figures 11 and 21); Drs. D. R. Gwadkin and S. K. Brandel and *Scientific American* (figures 12 and 37); The Department of Health and Social Security (figure 14); Dr. John Cairns, FRS, and W. H. Freeman & Company (table 3 and figure 15); Dr. John Cairns, FRS, and *Scientific American* (figure 18); Sir Richard Doll, FRS, Dr. R. Peto, and Oxford University Press (figures 16 and 17); Dr. M. F. Steward and G. P. Thomas Publishers (figures 19 and 22); Drs. L. G. Grabowski, J. M. Vernon, and L. G. Thomas and *The Journal of Law and Economics* (figure 20); Professor M. Schär (figure 23); U. S. Department of Energy (figure 25); Sir George Porter, PRS (figure 26); Sir Peter Baxendell and the Institute of Mining and Metallurgy (figure 27); *Nature* (figure 29); Lord Marshall, FRS, and *Atom* (figure 30); Prof. Guido Biscontin, Luigi Cattalini, and *Chemistry in Britain* (figure 31); Dr. Roger Revelle and

Scientific American (figure 32); Dr. W. S. Moore and Clarendon Press (figure 33); Dr. M. S. Swaminathan, FRS, and the World Meteorological Organisation (figures 34 and 35); The World Bank and Oxford University Press (tables 6 and 7, figure 38); the CIBA Foundation (table 4); Dr. R. L. Brown and *Science* (table 7).

I should like to thank the following publications for permission to reproduce articles that were originally published in slightly different forms:

"Is Science Necessary?" was originally published in German by Wissenschaftliche Verlagsgesellschaft, mbH, Stuttgart, in 1988 under the title *Ging's Ohne Forschung Besser?*

"Enemy Alien" was originally published in *The New Yorker* in August 1985 under the title "That Was the War: Enemy Alien." Copyright © 1985 M. F. Perutz. Reprinted by permission.

Uberto Limentani's account of the sinking of the *Arandora Star*, translated into English by D.J.H. Murphy, was published under the title "Survival at Sea" in the *Magdalene College Magazine & Record*, Cambridge.

"Atom Spy" was published in the *London Review of Books* in June 1987 under the title "Spying Made Easy."

"Discoverers of Penicillin" was published in the *New York Review of Books* in March 1986 under the title "Lucky Alec."

"Discoverer of the Atomic Nucleus" was published in the *London Review of Books* in June 1985 under the title "Lab Lib."

"Discoverer of the Quantum" was published in the *London Review of Books* in January 1987 under the title "German Scientist."

"Discoverers of the Double Helix" was published in *The Daily Telegraph* [London] in April 1987 under the title "How the Secret of Life Itself Was Discovered."

"Chemist into Statesman" was published in *Nature* (302) in 1983 under the title "Jewish Nationalism and the Liberal Ideal."

"How to Become a Scientist" was published in the *London Review of Books* in March 1981 under the title "True Science."

"Brave New World" was published in the *New York Review of Books* in September 1985 under the same title.

"Nature's Tinkering" was published in the *Times Literary Supplement* in September 1982 under the title "Nature's Bits and Pieces."

"Darwin, Popper, and Evolution" was published in *New Scientist* in October 1986 under the title "A New View of Darwinism."

"Department of Defense" was published in the *New York Review of Books* in October 1987 under the same title.

"More About Immunity" was published in the *London Review of Books* in September 1981 under the title "Iron Lady."

"Physics and the Riddle of Life" was published in 1987 by Cambridge University Press in *Schrödinger: Centenary Celebration of a Polymath*, edited by J. W. Kilmister. It was also published in *Nature* (326) in 1987 under the same title.

I should also like to thank the following for material and criticism: Dr. Douglas Bell, FRS; Sir Douglas Black, FRCP; Sir Hermann Bondi, FRS; Sir Arnold Burgen, FRS; Sir John Butterfield, FRCP; Dr. John Cairns, FRS; Professor Carlos Chagas, president of the Pontifical Academy of Sciences; Dr. R. J. Eden; Sir William Henderson, FRS; Dr. J. S. Heslop-Harrison; Sir Hans Kornberg, FRS; Professor Peter Lachmann, FRS; Professor Heinz Maier-Leibniz; Dr. John North; Professor Evamarie Sander; Mr. I. M. Sturgess; Dr. K. R. Williams; Professor E. A. Wrigley; and my daughter, Vivien Perutz. Finally, I thank Mrs. Christine Strachan for her accurate and fast typing of the many revisions of the manuscript.

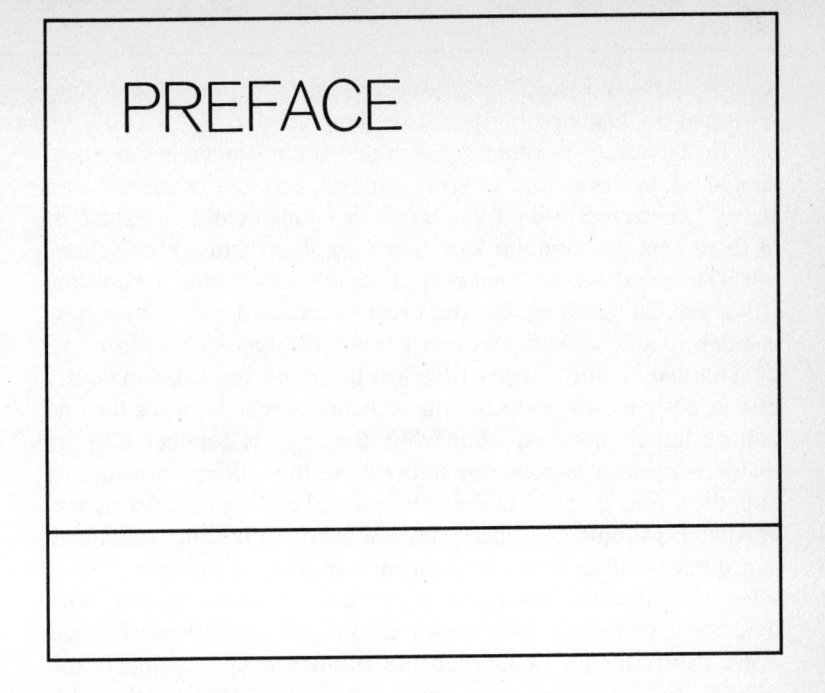

PREFACE

I am a molecular biologist who has spent his life finding out what the large molecules that are the workhorses of the living cell look like and how they function. One day, someone asked me to give a popular lecture on the impact that molecular biology has had on society, but this subject seemed to me premature. Instead, I talked about the impact that science as a whole has had on society, choosing three subjects that are basic to our existence—food production, health, and energy—and presenting them from a worldwide perspective of the past and the likely future. The first essay in this book, "Is Science Necessary?", has grown out of that lecture.

Scientists have changed our way of life more drastically than television stars, statesmen, and generals, but the public knows little about them beyond the caricature of the soulless hermit toiling away at abstruse problems that he cannot explain except in incomprehensible gibberish. The late Peter Medawar demolished this stereotype when he wrote, "It is high time that laymen abandoned the misleading belief that scientific enquiry is a cold dispassionate enterprise, bleached of imaginative qualities, and that a scientist is a man who turns the handle of discovery; for at every level of endeavour scientific research is a passionate undertaking and the Promotion of Natural Knowledge depends above all on a sortée into what can be imagined but is not yet known."[1]

In science, as in other fields of endeavor, one finds saints and charlatans, warriors and monks, geniuses and cranks, tyrants and slaves, benefactors and misers, but there is one quality that the best of them have in common, one that they share with great writers, musicians, and artists: creativity. People's minds prefer traveling along familiar grooves, and the creation of anything entirely new is indescribably difficult (indescribably in Wittgenstein's sense), so difficult that no one can describe how it is done. Imagination comes first in both artistic and scientific creation—which makes for one culture rather than two—but while the artist is confined only by the prescriptions imposed by himself and the culture surrounding him, the scientist has Nature and his critical colleagues always looking over his shoulder. To paraphrase Winston Churchill: "In science you don't need to be polite, you merely have to be right." Great scientists and artists have one other trait in common: they both tend to be single-mindedly devoted to their work. Renoir painted every day of his life, and when old age had made his fingers too arthritic to hold a brush, he got someone to tie the brush to his hand. Haydn rose early each morning to compose; if ideas failed him, he clasped his rosary and prayed until Heaven sent him fresh inspiration. Tolstoy rewrote *War and Peace* seven times. When Newton was asked how he had arrived at his insights, he answered, "By keeping the problem constantly before my mind."

There is little benefit in following scientists' daily grind but much in tracing the unique combinations of theoretical knowledge and manual skills, the web of personal encounters and accidental observations, the experience, temperament, moods, and clashes that go into the making of discoveries, even though the crucial leap of the mind is often impenetrable. There is also something to be said for finding out why others, seemingly just as able, were too blind to grasp what Nature tried to tell them.

True science thrives best in glass houses, where everyone can look in. When the windows are blacked out, as in war, the weeds take over; when secrecy muffles criticism, charlatans and cranks flourish. One of my essays describes my exertions on a wartime project that belongs to the realm of science fiction, and another deals with the split personality of an atom spy.

When debating public issues that touch on commercial matters, members of the British Parliament are expected to declare their interest. Readers of my essay "Is Science Necessary?" may suspect me of being in the pay of companies making agrochemicals, drugs, genes, or nuclear power plants, but I have no vested interests in any of these. My sole interest is in the survival of Nature and of civilization.

When writing about science, I have a parrot perched on my shoulder, calling out every so often, in his raucous voice: "Can't this be said more simply?" I hope that his shrill admonitions have helped me to make these essays intelligible to the layman.

IS
SCIENCE
NECESSARY?

IS SCIENCE NECESSARY?

THE HUMANIZING INFLUENCE OF SCIENCE

Is scientific research the noblest pursuit of the human mind, from which springs a never-ceasing stream of beneficial discoveries, or is it a sorcerer's broom that threatens us all with destruction? Has science spoiled the quality of life?

You need only go back to your grandmothers' time to realize that Adam's eating from the apple of knowledge has been of the greatest benefit to Eve. Remember the beginning of *Anna Karenina*: at the Oblonskis' house everything is topsy-turvy because the princess has found out that the prince has been having an affair with their French governess. What drove him into her arms? Dolly, the princess, is only thirty-three, and already the mother of five living and two dead children. Her many pregnancies have left her faded and plain, making the prince lose interest in her. Even upper-class women in nineteenth-century Europe received little education, and their role was confined to childbearing and housekeeping. Many women died after childbirth of puerperal fever, an infection preventable by simple hygiene and therefore almost forgotten today. For working-class girls who were unmarriageable because they lacked a dowry, domestic service was the only outlet. The National Gallery of Scotland has a charming picture by Gerard David, dating from about 1500, in which St. Nicholas surreptitiously drops a moneybag into the bedroom of his bankrupt friend's daughters, so that his friend can buy them husbands. Women's liberation could not have succeeded if science had not provided women with contraception and household technology.

All former civilizations not only were male dominated but were also based on slavery. This was true, of course, of Greece and Rome, but few people realize that it was also true in the Italian Renaissance. In 1395 Francesco Datini, a merchant in Prato, a small town near Florence, wrote to his partner in Genoa, "Pray buy me a little slave girl, between eight and ten years old, and she must be of good stock"—as if she were a horse.[1] Even in the eighteenth century most servants were free only in name. In *The Marriage of Figaro*, Figaro and Susanna were able to outwit Count Almaviva, but they could never have escaped him.

Science has humanized society in other ways, but the process has been a very gradual one. The burning of witches reached its peak in the seventeenth century, the time of Galileo and Newton, and ceased only in the eighteenth.[2] In eighteenth- and early-nineteenth-century England, over two hundred offenses were punishable by death. In an account of a group of young boys whom a judge condemned to death, the witness wrote, "Never have I heard boys cry so." Dr. Samuel Johnson, the lexicographer who is celebrated as one of the most enlightened men of eighteenth-century England, used to entertain himself and his friends on a Sunday by watching the lunatics chained up in Bedlam (figure 1). In my own youth jokes about lunatics were still common.

What has changed our attitude toward wrongdoers and the mentally sick is a combination of science and humane liberalism that asks, "Is hanging an effective deterrent? Are madmen and demented old women possessed by the devil? What causes madness and crime?" Few countries have reason to be proud of their prisons and mental hospitals, but science has changed our attitudes toward human behavior, gradually substituting reason for cruelty, prejudice, and superstition. This approach has grown slowly and needs to be preached anew to every generation. Otherwise, it is only people's bodies that are jet-propelled while their minds revert to the Middle Ages.

People quickly get used to science's technical achievements

1. Bedlam, the London lunatic asylum, in 1734. Engraving by William Hogarth.

while ignoring its laws. According to Martin Gardner,[3] President Reagan regularly consulted astrologers before making major decisions, unaware, it seems, that nearly 1,600 years ago Saint Augustine wrote in Book V of the *City of God*:

> How comes it that astrologers have never been able to assign any cause why, in the life of twins, in their actions, in the events which befall them in their professions, arts, honours, and other things pertaining to human life, also in their very death, there is often so great a difference, that, as far as these things are concerned, many entire strangers are more like them than they are like each other, though separated at birth by the smallest interval of time, but at conception generated by the same act of copulation, and at the same moment?

Recently several fellows of my Cambridge college believed that the spoon-bending magician Uri Geller could invalidate the laws of physics.

When we come to the condition of the common man, there is a great difference between the approach of the priest, of the politician, and of the scientist. The priest persuades humble people to endure their hard lot; the politician urges them to rebel against it; and the scientist thinks of a method that does away with the hard lot altogether. Thus, science has brought about the kingdom of freedom of which Karl Marx wrote: "It begins where drudgery ends." In some parts of the world, notably Scandinavia, Austria, and New Zealand, there are now no longer any crass contrasts between rich and poor, and the Christian ideal of the equality of man has at least been approached. In these countries Marx's dictum that we "can buy a greater degree of freedom only by enslaving other men" has been disproved by science. A ruling class bound to oppress and fight the ruled no longer exists, and political power has ceased to be "the organized power of one class oppressing the other." There is no dictatorship either of the bourgeoisie or of the proletariat, because science and democratic socialism have lifted the standard of living of the masses to a level that in Marx's time was hardly dreamed of by the bourgeoisie.

Poverty was the greatest and most intractable social problem of eighteenth-century Europe. In Munich, the poor lived on the streets or herded together in terrible slums. Diseased beggars in rags were everywhere, and people paid them quickly to get rid of them. Conditions resembled those of Calcutta today, except that in Calcutta at least the poor don't freeze.[4] London was little different: the German scholar Georg Lichtenberg complained of not being able to go around without being pestered all the time by prostitutes and pickpockets, many of them children.[5] In the countryside, bad harvests and severe winters often decimated the population. Science and technology have eliminated such misery in much of the world today.

How did the myth of the simple, blissful, and harmonious life of the past originate? Poets of antiquity loved to conjure up a rustic Arcadia. In the eighteenth century, when the poverty and squalor of country life were common knowledge, neither Boucher's delectable pastoral scenes nor Marie Antoinette's farm deceived those who relished them, but in the nineteenth century, thanks to the Arts and Crafts movement, the myth began to color the lives of thousands who wanted to escape from the ugly world of technology to a simple rustic existence that was healthy and morally good. Heirs to William Morris's clients, or to those who, in the 1880s, sought village bliss within the confines of London and Bedford Park, are today's Arcadian seekers, who make a cult of health foods, frequent herbalists, dress in romantically rustic floral prints, buy pinewood furniture for their suburban cottages, or turn to organic farming. Are they aware that they have built an ancient Greek myth into their lives? Could the desire to escape into that myth have engendered the antiscience sentiment so prevalent today? This sentiment is played on by cranks and exploited by publicists skilled in misrepresentation.

CHALLENGES FOR SCIENCE

Yet have we not had the best of science? Have we not come up against the law of diminishing returns, ever larger expenditures being needed for ever smaller advances? Would it not be better to call a halt to research and get along with existing knowledge, using the money saved to reduce taxes? This experiment was tried in China by what has been euphemistically called the Cultural Revolution. Scientists were harnessed to the plow; research institutes were closed or their work was paralyzed by perpetual discussions of its political aims. Self-seeking scientists were ordered to shed their narcissistic images and publish their work anonymously, attributing their success solely to the wise guidance of Chairman Mao.

What was the result? Did the Cultural Revolution lead Chinese people back to the ideal of Rousseau, the ideal of so many young people in the West today, a society of noble men and women in harmony with Nature? *No*, it brought them to the brink of economic collapse, because the problems of keeping everyone fed, clothed, and in reasonable health and of protecting the country from foreign invasion cannot be solved without science. This is not merely because new problems perpetually cry out for solution but because existing knowledge cannot be applied intelligently; problems cannot even be formulated without advanced scientific training. So science is here to stay; we cannot wish it away but must use it to the best advantage. There exists, however, a fundamental dilemma that scientists and society find difficult to face.

Science often exacts a price. Most technical advances are subject to Niels Bohr's principle of complementarity, which he formulated to explain that waves and particles are dual aspects of matter. According to this principle, benefits and risks are complementary aspects of each technical advance. Society must judge between them, but such judgment can present us with agonizing choices where neither moral values nor scientific facts lead us to clear decisions.

For instance, civilization demands that each human being has a right to expect a reasonable span of life free from hunger and disease. Even partial fulfillment of this expectation has given rise to an exponential growth of population that threatens to defeat the very demand that produced it.

The replacement of slaves by machines requires energy, whose consumption at an ever-increasing rate threatens to destroy the civilized life it is supposed to sustain.

Civilized society will survive only in conditions of national and international peace, while science puts into its hands ever more elaborate, costly, and effective means for its own destruction. These three challenges are interrelated, but I shall discuss them in turn.

SCIENCE AND FOOD PRODUCTION

AGRICULTURAL YIELDS

In *Gulliver's Travels* Jonathan Swift wrote of the king of Brobdingnag that "he gave it for his opinion that whoever could make two ears of corn, or two blades of grass to grow upon a spot of ground where only one grew before, would deserve better of mankind and do more essential service to his country than the whole race of politicians put together. . . . " Yet I have seen no monuments erected to Norman Borlaug, the American who developed high-yielding wheat, nor to Douglas Bell, the Englishman who developed high-yielding barley. Their names are unknown to the great public, and only the shortcomings of these high-yielding varieties are broadcast.

Science has revolutionized agriculture, doubling the world's grain production from 1950 to 1971, but can it continue to feed the world's growing population without unacceptable damage to the environment? To answer this question, let me first report on progress in one industrialized country, Great Britain, and one developing country, India. In Great Britain spectacular increases in agricultural productivity have been achieved by a combination of botany, genetics, chemistry, and engineering, stimulated by guaranteed prices that were above prices on the world market (figure 2). In the 1930s Britain produced only one-third of its food; now it produces 80 percent for a larger and better fed population with fewer farm workers on less land. In addition Britain now exports about $3 billion worth of farm products. The introduction of tractors has freed about 10 million hectares on which farmers had to

Annual average milk yield per dairy cow in England and Wales

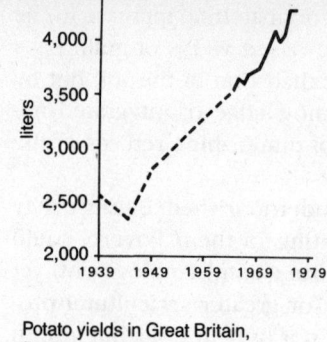

Wheat yields in Great Britain, five-year averages

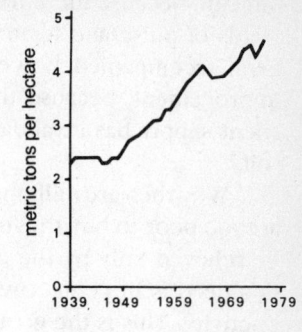

Potato yields in Great Britain, five-year averages

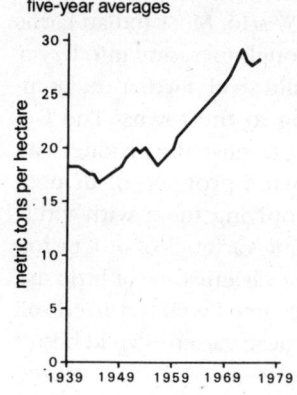

2. Typical rises in British agricultural yields, 1939–1979. SOURCE: "Dairy Facts and Figures," by the Milk Marketing Board and Ministry of Agriculture, Food and Fisheries as reproduced by Her Majesty's Stationery Office in *Seventh Report of the Royal Commission for Environmental Pollution, Agriculture, and the Environment*, Cmd. No. 7644, 1980.

grow food for their horses. British yields of wheat per hectare are still rising at the rate of about 2 percent per annum. A cow now gives nearly twice as much milk a day, grazing on a smaller field, than her great-grandmother did in 1946. Yet productivity could be raised much further. Even now, average crops are no more than half those obtained by some good farmers, and these in turn are lower than those that can be obtained in experimental work.[6] British achievements in agriculture are typical of those of other Western democracies.

For a decade after the end of World War II, it looked as though India would be overtaken by a Malthusian catastrophe, but so far this has not happened. Between 1945 and 1981 India's population rose from 300 to 750 million, but the country's amount of edible cereal per head has increased steadily: India now produces enough grain for her entire population and can even build up reserves against droughts and floods. India was able to send grain for the relief of the famine in Kampuchea. Production of protein is more difficult, because there has as yet been no matching increase in the yields of pulse and legume crops. The raised yields of grain have been accompanied, however, not by exhaustion of the soil but by improvement, because the growing knowledge of integrated nutrient supply has invalidated the law of diminishing returns of the soil.[7]

Why then are millions of Indians undernourished? Because they are too poor to buy the food that is waiting for them. Poverty could be relieved only by the creation of more gainful employment, yet this need is in conflict with the need for greater agricultural productivity. This is the greatest problem not only in India but also in Italy and in many countries in the Third World. Most Indian farms are smaller than half a hectare, but their conglomeration into larger units, to make them more productive, would swell further the number of unemployed rural workers drifting to the towns. The Department of Agriculture is therefore trying to raise the productivity of small farmers by means of a countrywide program of demonstration and support, for instance, by supplying them with virus-free seed potatoes and seeds of high-yielding varieties of other crop plants. It is often alleged that high-yielding varieties are of little use to underdeveloped countries, because they need well-fertilized soil and are susceptible to local pests. In fact, these varieties yield better ratios of edible starch to inedible cellulose even on poorly fertilized soil, and their susceptibility to local pests can be overcome by hybridization with local disease-resistant stocks.

What of the future? In India scientific methods exist for raising food production to feed a doubled population in twenty years' time, but will they be applied? M. S. Swaminathan, the architect of modern agricultural planning in India, has written, "The most urgent task is to generate the social collaboration and the requisite blend of political will and professional skill for converting agricultural assets into wealth meaningful to the people." Swaminathan has shown that productivity of the land can be raised not by any single grandiose plan but by intelligent scientific attention to a thousand important details. However, exponential population growth cannot be matched by rising agricultural productivity indefinitely.[8]

RICE

Most of the populations of East Asia live mainly on rice. Rice was planted there as early as seven thousand years ago, and for

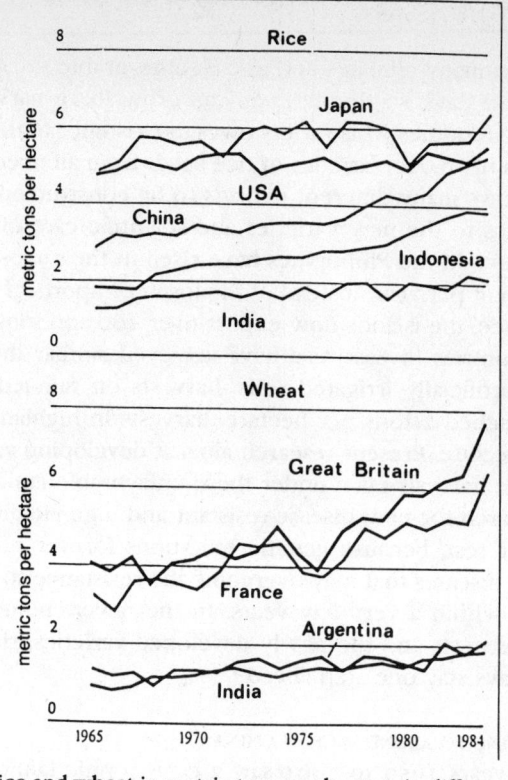

3. Yields of rice and wheat in metric tons per hectare in different countries from 1965 to 1984. Note the astonishingly steep rise of wheat harvests in Western Europe. SOURCE: Food and Agriculture Organization.

centuries it has been planted every year in the same fields. Modern research has shown that exhaustion of the soil was prevented by ferns, algae, and bacteria that live in the rice paddies and fix about 30 kilograms of nitrogen per hectare from the air; their activity provides enough fertilizer for harvests of 1 to 2 metric tons per hectare. The new varieties of rice have raised yields in Japan, Korea, Australia, and America to about 6 tons per hectare; an outstandingly able Japanese farmer has even raised 12 tons per hectare. Such large harvests need extra nitrogen that must be supplied by chemical fertilizers (figure 3).

The first high-yielding varieties of rice suffered from sensitivity to pests and diseases. To meet these challenges, the International Rice Research Institute in the Philippines has developed a variety that is resistant to four frequent insect pests and five serious diseases; this variety has already been planted on 10 million hectares.

It grows in many climates and also in unfavorable soils; it matures in only 110 days, so that farmers can grow three harvests a year on irrigated paddies. This variety owes its existence to the institute's collection of 70,000 varieties of rice seeds from all over the world, which allows many different hybrids to be constructed.

Thanks to the new varieties and scientific care of plants and soil, harvests in the Philippines have risen in the 1970s and 1980s by 5 percent per year; instead of the former imports of large quantities of rice, the islands now export over 100,000 tons each year. Other countries in East Asia have achieved similar increases but only on artificially irrigated soils; harvests on rain-fed soils have hardly reached 2 tons per hectare, harvests in highlands barely 1 ton per hectare. Present research aims at developing varieties that yield large harvests even under these unfavorable conditions.

Research for new disease-resistant and high-yielding varieties can never rest, because genetic mutations forever produce new pests and diseases that may overcome the resistance so laboriously achieved within a very few years. In the never-ending battle between the pests and the newly developed varieties, rice research must always stay one step ahead.[9]

FROM FAMINE TO ABUNDANCE IN CHINA

The years 1959 to 1961 saw a catastrophic famine in China; its causes might have served as a warning to the leaders of other countries if it had not been kept secret until recently. It is now known that this famine cost at least 16 million and perhaps as many as 30 million lives; the children whom it left physically and mentally crippled have never been counted. The famine was caused only in small part by barrenness of the soil resulting from droughts; in large part it was caused by the barrenness of Marxist thought. Mao's Great Leap Forward abolished private markets in favor of collectivized farming. Production, storage, trade, and consumption of food all had to work according to one central plan. Provinces had to be self-sufficient in food so as to minimize trade. Since every official had to report that his part of the central plan had been accomplished, the sum of all the reports showed that from 1957 to 1958 the grain harvest had nearly doubled (from 195 to 375 million tons); the government thereupon decided to reduce cereal production in 1959 by 5 percent in order to free land for the production of industrial raw materials. All the same reports showed that the grain harvest again yielded 375 million tons, exactly as planned. In the autumn of 1958 two members of the Politburo actually traveled to two provinces and discovered that the harvests were lower than had been reported, but this discovery cost them their jobs. Only after Mao's death did the authorities admit that in 1957 the real harvest had amounted to only 170 million tons and that it had dropped to 143 million tons in 1960 (figure 4). In the country, people starved to death because the little food they had was taken away from them to feed the towns. After 1960 China began to import cereals and pay farmers better prices, which gave them an incentive to raise their harvests, but by 1978 the cereals produced per head of population were still no more than in 1952. After Mao's death a combination of scientific agriculture and political reform brought a rapid recovery; cereal harvests rose annually by 7 percent (truly this time) and productivity per farm worker by 12 percent, so that China's enormous population is now fed better than ever before (figure 5). There is concern, however, that in the last thirty years China lost a tenth of its agricultural land through building,

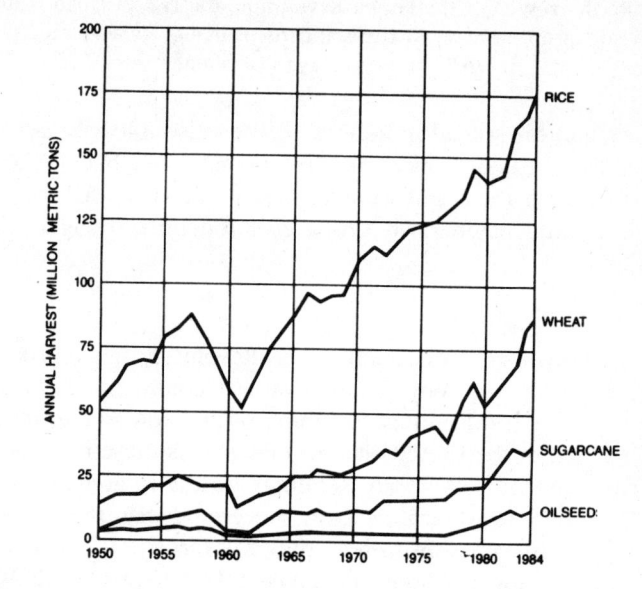

4. Harvests in China, 1950–1984. The catastrophic fall between 1957 and 1961 accompanied Mao's Great Leap Forward. The recent steep rises result partly from the Green Revolution and partly from a return to private enterprise. SOURCE: Vaclav Smil, "China's Food," *Scientific American* 253 (December 1985): 104.

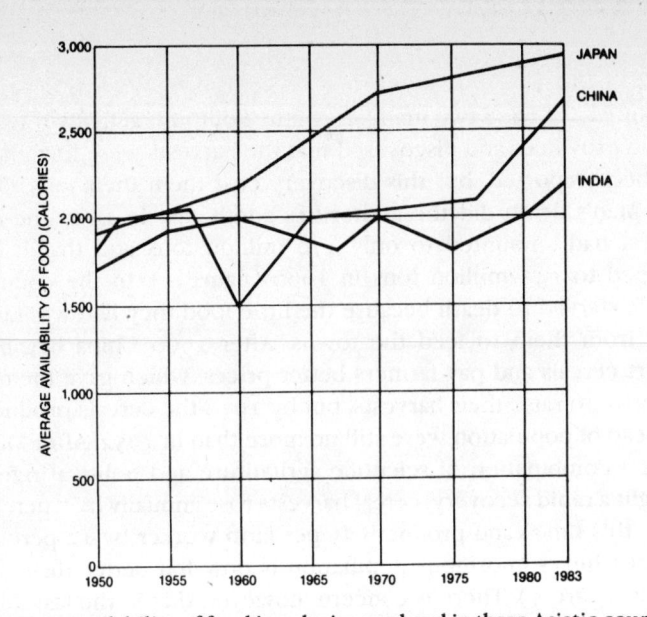

5. Average availability of food in calories per head in three Asiatic countries, 1950–1983. The striking rises in China and India after 1975 are in a large measure the results of the Green Revolution; the rise in Japan is mainly the result of an increase in the standard of living. SOURCE: Vaclav Smil, "China's Food," *Scientific American* 253 (December 1985): 104.

erosion, and the spread of deserts, so the cultivable area per head of population has by now shrunk to one-tenth of a hectare. Szechwan has lost a third and Yunnan nearly half of its forests. Even today's central planners in China give too little thought to the future.[10]

STAGNATION IN AFRICA

Why was Asia's successful Green Revolution not reproduced in Africa (table 1)? Why is much of that continent's population undernourished and subject to famines? The Food and Agriculture Organization of the United Nations (FAO) lists these reasons:

Government policies are generally hostile to agriculture. For example, if farmers grow more food than they need for their own subsistence, they are forced to sell it to the state for less money than it cost them to produce it. While West European and North American governments subsidize their agriculture, African governments use agriculture as a source of taxation that expropriates the farmers.

Populations are increasing fast.

Cultivation of new land is slowing down.
Technical backwardness makes yields stagnate or even drop.
Deserts are spreading, and soils are becoming exhausted and
    salty.
Global economic developments render it increasingly difficult
    for African states to balance their budgets.

In many African countries agriculture accounts for over half
of production and employment, but governments spend on it less
than a tenth of their incomes. Peasants receive no support for the
growth of food for home markets, and the prices of their crops are
undercut by cheap imports, because the many shaky governments
have to satisfy the townspeople's demand for cheap food. Research
and fertilizers, if available, are reserved for export crops. Consequently peasants lack incentives to grow more food than they need
for their own subsistence. The Ethiopian famine in 1985 shocked
the world, but many African countries are approaching more severe
famines, because their populations are increasing by 3 to 4 percent
per year while their agriculture stagnates, and their earnings of
foreign currency are insufficient to import enough food for their
people. If these developments were to continue unchecked, agricultural deficits, especially in countries south of the Sahara, will
exceed anything coverable by commercial imports and foreign aid.
The FAO estimates that at constant prices African imports in 2010
would cost $2.2 billion, while the continent's agricultural exports

TABLE I

CHANGES IN WHEAT PRODUCTION
IN SELECTED ASIAN AND AFRICAN COUNTRIES,
1971–1984

	AVERAGE % CHANGE PER YEAR
Indonesia	5.2
Korea	5.0
Pakistan	4.3
Gambia	− 0.3
Zambia	− 2.2
Ghana	− 2.4

SOURCE: *African Agriculture: The Next Twenty-five Years* (Rome: Food and Agriculture Organization, 1986).

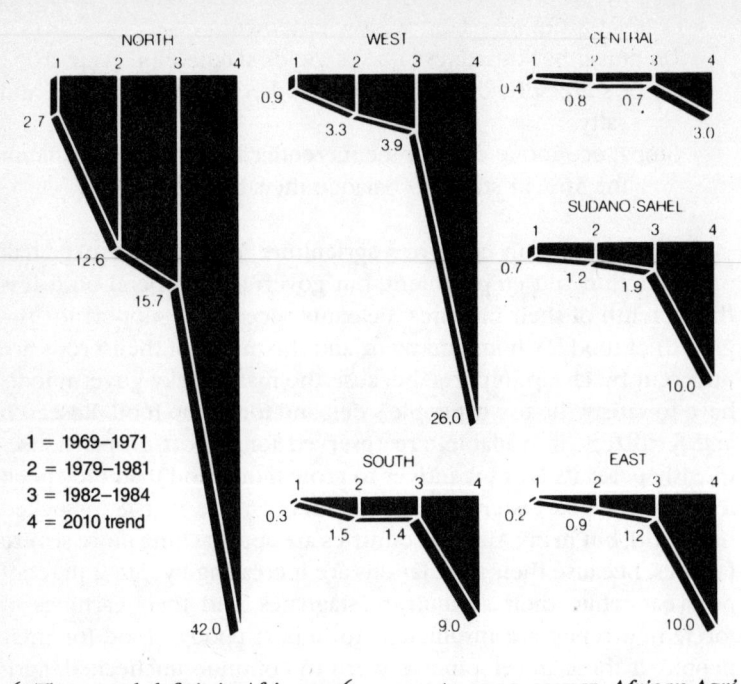

1 = 1969–1971
2 = 1979–1981
3 = 1982–1984
4 = 2010 trend

6. The cereal deficit in Africa, 1969–1971 to 2010. SOURCE: *African Agriculture: The Next Twenty-five Years* (Rome: Food and Agriculture Organization, 1986).

would yield maximally $1.4 billion. The annual cereal deficit alone would amount to 100 million metric tons, compared with a present annual world surplus of 12 million tons (figure 6).

African soils are often alleged to be too infertile for large yields, but the FAO has shown that this allegation applies to only a few countries where soils are too dry and cannot be irrigated. In most countries agricultural production could be doubled by the application of existing technology, given the will, the incentives, and, above all, the necessary adaptive research. The FAO has demonstrated the feasibility of such improvements by a series of successful projects, but their implementation would require radical changes in the countries' economic policies and could not be achieved without foreign aid—not for grandiose projects like the Aswan Dam, which has proved a disaster, but for local projects of irrigation, improvement of seeds and soils, desert containment, and pest control elaborated with local farmers and supported by research on the spot. African countries need heat-, salt-, and drought-resistant varieties of wheat; rice that matures fast at high altitudes; rice that is resistant to aluminum in the soil; and cereals that are resistant to the pests prevalent in different regions. The FAO emphasizes that these problems can be solved only by well-supported and competently staffed *local* research institutes, of which only few exist, because African states have been indifferent to such needs.[11]

Fifty years ago only 9.5 million people lived in Ethiopia, and half the country was wooded. By the 1980s Ethiopia's population had grown to 42 million, and only a third of its forests remained. As trees were cut down, the soil was washed away from the dead roots; the environment began to change, and the desert moved in. A prolonged drought aggravated the damage done by man. These were the origins of the great recent famine, but even so it could apparently have been averted. In the early 1980s an international commission, appointed jointly by two agencies of the United Nations and the Ethiopian government, examined the country's agriculture. It reported that agricultural production had been dropping by 5 percent per year even before the long drought and that rural incomes and productivity had sunk too low to finance agricultural and general economic development. The commission recommended that surplus farm workers be used to bring more land under the plow, to raise yields by means of small local irrigation plants, to improve farm tools, and to build small local industries for the processing of farm products and the production of consumer goods. It further suggested that these enterprises not be organized in collectives, which have to deliver all their products to the state, but in local cooperatives that invest their profits in their own workers. The commission wrote,

> Great care must be taken to ensure that cooperation is in the material interest of the peasantry. Two dangers in particular must be avoided. First, force must not be used to establish cooperatives. It will only lead to resistance; it will not lead to accumulation.
>
> Second, the incomes generated by cooperatives must not be appropriated by the state in the form of high taxes or compulsory deliveries of grain at fixed low prices. . . .
>
> Ethiopia has systematically under-invested over a long pe-

riod in rural infrastructure—roads, power, irrigation, storage and processing facilities—and in the health, education and training of the rural population. . . .

They should be encouraged to expand their activities to include health, nutrition, family planning and the care of children.

The Ethiopian government rejected these recommendations and even succeeded in suppressing publication of the commission's report by the United Nations agencies that had sponsored it (fortunately, it did not prevent publication of the report's summary in the English newspaper *The Guardian* two years later).[12] The reasons for Ethiopia's rejection of the report's recommendations may be sought in the Marxist dogma that industrialization is the first step toward economic growth and that capitalism must be suppressed by organizing agriculture in collectives that are run like factories. Politically, rural misery hardly matters as long as government supporters in the towns are satisfied with cheap food. Similar attitudes held by the dictatorships of the left and right also oppose the FAO's suggested reforms in other African countries.

Economists of the World Bank have shown that, contrary to Marxist teaching, increased agricultural production is the first step toward industrialization and economic growth. They quote many years' experience showing that agricultural health is essential for national growth and that, in the long run, taxation of agriculture in order to force industrialization damages both. With the exception of oil and mineral exporters, countries with large agricultural growth also enjoy great industrial growth. Both in England and in Japan agricultural growth preceded industrialization, because it generated the necessary capital and demand for consumer goods.

I was sorry to note that the FAO's publications gloss over the most urgent of all Africa's problems: the population explosion that threatens to double the number of people in many countries in hardly more than twenty years (eighteen years in Kenya), and I have heard of no measures taken by African countries themselves to halt that explosion. This state of affairs will lead to catastrophic famines unless the African population is decimated by the fast-spreading AIDS epidemic (in Zaire 10 percent of the population carry the virus).

In earlier times famines often occurred in Europe; later a famine was reported in some part of the world almost annually. For example, in India more than 10 million people starved to death in the 1770s and in the 1860s and in China similar numbers died in the 1870s. Since 1940 there have been worldwide about a dozen famines, but most of them were less extensive than those in former centuries and several were caused by wars. Thanks to the growth of international trade in cereals, countries are able to import grain in emergencies, and they also have better means of transporting it to hungry areas. In former times international aid on the present scale did not exist. All the same, the World Bank estimates that globally the undernourished number between 340 and 730 million; this excludes China. The World Bank takes the view that undernourishment is generally caused not so much by shortage of food as by poverty and maldistribution of income. The best remedy is economic growth.[13]

RESEARCH AND THE SMALL FARMER

Over 90 percent of the world's agriculture is in the hands of small farmers, who are often alleged not to have benefited from the Green Revolution, but the only hope of raising agricultural output to keep in step with population growth rests on raising small farmers' productivity. Experience has shown that this is also the best way to prosperity. The World Bank has therefore tried to devise a system that takes the problems of individual small farmers to research institutes and carries the results back to the farmers themselves. The system originated from expert agricultural advice on the European pattern, but this proved unsuitable for developing countries, because it generated no exchange of information or trust between farmers in the country and research institutes in town. Information only went from the institutes to the farmer; his problems were not brought back to the institutes, which optimized harvests under their own artificial conditions without troubling to find out why small farmers failed to match their successes.

In 1977 the World Bank formulated its Training and Visiting Scheme, which was tried first in Turkey and then introduced into several provinces of India. The scheme is based on close collaboration among farming families, agricultural extension workers, scientists, and administrators and is aimed at combining traditional and scientific knowledge to give farmers applicable, clear, and sensible advice. At the bottom of the ladder is the village extension worker, who must visit selected contact farmers at least every two weeks and provide them with three or four pieces of advice for their work during the succeeding two weeks. These contact farmers must pass the advice on to others, so that one extension worker's advice reaches between five hundred and twelve hundred farming families. In any two-week period the extension worker spends eight days on such visits and one day on a teaching session with a specialist, who discusses the problems encountered by the farmers and provides advice to pass on to them for the next two weeks. The specialist has an equally structured program, which divides his time between teaching sessions, research, and work in the field. From him the line of command leads back to the administrators and research institutes.

Such a hierarchy sounds military in its rigidity, but appears to be necessary for effective collaboration of this sort. Advice aims at introducing cheap methods based on scientific research that are applicable to farms of any size and likely to raise their incomes above those attainable by traditional methods alone. Its principal targets are improved seeds and better control of pests, weeds, and water management, all adapted to the local farming system. For example, in one Indian province the yields of wheat declined despite optimal fertilization and irrigation. The decline was found to be caused by a shortage of zinc in the soil, which could be cheaply remedied. That was a simple problem, but others can be much harder to solve and need first-class, patient, and devoted research.

The weaknesses of this highly organized system appear to be caused more by human than by scientific failings. Scientists in research institutes want to do fundamental work that leads to publication, on which their careers depend, but such work often lacks practical objectives. The scientists get more pay and belong to a higher social class than the village extension workers, who in turn make farmers feel their superiority and tend to lecture them rather than listen to them. This problem of class snobbery can sometimes be remedied by farmers choosing one of their own members for training as an extension worker, and also contributing to his pay.

Despite these weaknesses, this Training and Visiting Scheme seems to be the most effective yet devised for raising small farmers' yields and incomes. The Indonesian government was quick to recognize this and by 1983 already employed 15,000 village extension workers, mainly in order to raise rice harvests. In Thailand farmers were advised to grow cassava in addition to rice. They soon found out that doing so paid, especially since the government supplied them with free seed and fertilizer. As a result the annual production of cassava rose from 1 to 12 million tons within seven years. The system helped about 100,000 farmers in Upper Volta to improve their harvests of cotton and food sufficiently to afford better seeds and more fertilizer. It is also being applied successfully in India, Bangladesh, Pakistan, Nepal, Sri Lanka, the Philippines, and several African states, but in Africa governments are reported to have delivered the seed only after the sowing season was past.[14]

FERTILIZERS

In Britain the use of phosphate fertilizer has increased threefold, that of potassium tenfold, and that of nitrogen thirtyfold in the past fifty years. The world's consumption of phosphate is rising at an annual rate of 6 percent, but there is no danger of exhaustion, since several countries possess enormous reserves of phosphate rock. The richest are in Morocco.[15] Nitrogenous fertilizers are made from air and methane that supplies both the hydrogen and the energy needed to reduce the nitrogen to ammonia. Reserves of potassium are also plentiful. Hence there exists no danger of fertilizers becoming scarce as long as we have enough energy. In the 1970s developed countries used about 3.0 percent of their consumption of fossil energy for agriculture; about 0.7 percent was used for nitrogenous fertilizers, whose manufacture is a $10 billion business. Four times more energy (12 percent) is consumed for transport, processing, distribution, and cooling of food. In developing countries the fraction of total energy consumption needed for nitrogen fertilizers is much larger, and energy for cooking is also in short supply.

What are the risks of increased agricultural productivity? When I set out to write this, I believed that it had been achieved only at the cost of damaging environmental pollution by agrochemicals. I therefore examined the scientific evidence, especially that collected by the British Royal Commission for Environmental Pollution, which was composed of five scientists, two medical men, one engineer, and seven laypeople, including one trade unionist, none of them having vested interests in farming or agrochemicals.[16] Here is what I learned from their reports and from other literature.

Potassium is harmless. Phosphate is harmless to man, but not to all his environment. Phosphate runoff is believed to have killed life in the Great Lakes by promoting the growth of algae that shut out light and caused wide variations in the level of dissolved oxygen. The invertebrates that are the source of the food chain could not survive this, and as a result all other life died out.

The intensive application of nitrogen may raise the concentration of nitrate in drinking water above the safety level of 50 milligrams per liter recommended by the World Health Organization, which feared that higher levels might lead to an increased incidence of infantile methemoglobinemia (a blood disease) and possibly also of cancer of the gastrointestinal tract in adults. We have in fact been drinking water containing 50 to 100 milligrams nitrate per liter for years, and the nitrate level has risen to about 100 milligrams per liter in some areas of Britain, but so far there has been no evidence for raised incidence of any cancer in those areas. It might be argued that it could be twenty years before a higher incidence of cancer shows up, but in fact nitrate levels in these areas have been rising for many years, while the frequency of gastric cancer has decreased, and the area with the highest nitrate level in Britain has the lowest incidence of gastric cancer. The situation is being watched; if the nitrate level rises any higher, the Royal Commission recommends removing nitrate from drinking water rather than restricting the application of fertilizers.

Some friends of Nature want us to use animal manure instead of fertilizer to avoid these risks (they believe that such organic food would be healthier), but there is no scientific evidence for manure being cleaner than ammonia, and the amount of manure and other organic waste available in Britain is insufficient to sustain agricultural yields. Instead, scientists are trying to extend the range of plants capable of fixing nitrogen from the air.

Legumes such as soybeans do not require nitrogen fertilizers, since their root nodules contain bacteria that fix nitrogen from the air. The same is true of certain varieties of sugarcane.

In Brazil sugarcane is cultivated on a huge scale for the production of alcohol as a fuel. By 1985 alcohol production amounted to 200 million liters, and since then nearly all new cars in Brazil have been designed to use 95 percent alcohol instead of gasoline. Selection and inoculation with nitrogen-fixing bacteria has yielded a sugarcane that takes half of the nitrogen needed for its growth from the air, thus providing a positive energy balance (in other words, more calories are gained in the form of sugar than have been invested in the form of labor and fertilizers).[17] This is a spectacular success, but the fraction of naturally fixed nitrogen can probably be raised still further. In the Far East, scientists are trying to improve upon the centuries-old symbiosis of rice with the fern *Azolla* and the bacterium *Anabaena*, which grow on the surface of the rice paddies and fix nitrogen from the air. Plant breeding stations in several countries have been trying for some years to grow wheat, sorghum, or millet plants that will live in symbiosis with nitrogen-fixing bacteria, and such experiments have already produced promising results.

PESTICIDES

Crops are infected by insect-borne viruses, fungi, and worms, and crop plants must compete with weeds and avoid damage by insects and plant eaters. In modern agriculture these pests are kept in check by chemicals that are as vital as fertilizers for the maintenance of high yields; all the same, there is much public anxiety about them (figure 7).

Insecticides have acquired a bad name because DDT accumulated in certain food chains, causing populations of many wild birds to be reduced and some other creatures to be killed, and because DDT has become a persistent contaminant of our environment. Yet DDT is no more toxic to man than aspirin is and has killed people only when mistakenly eaten in place of flour. Not only has it protected crops from virus infection, but it has also eliminated plague and typhus from most of the world, and malaria from large parts of it (figure 8).

Malaria used to be endemic in many parts of Italy, and its incidence increased catastrophically during the Second World War, so that 400,000 cases per annum were registered in 1946–47, with a mortality rate of over 40 per 1,000 inhabitants at risk. Following an intensive DDT campaign, no indigenous case of *Plasmodium falciparum* has been reported in Italy since 1952. In 1946 there were 2.8 million cases of malaria in Sri Lanka, but after an efficient DDT spraying campaign only 100 cases were reported in 1961 and 150 in 1964. In that year spraying was stopped. Four years later there were 440,644 cases, and this figure rose to 1.5 million in 1970. Much the same happened, on a larger scale, in India. If these countries had persevered with the spraying, as Italy did, they might have rid themselves of malaria. As it turned out, they missed their chance, and by now the malaria-carrying mosquitoes have become resistant to DDT and most other insecticides. Environmentalists who decry the use of pesticides must weigh these figures against

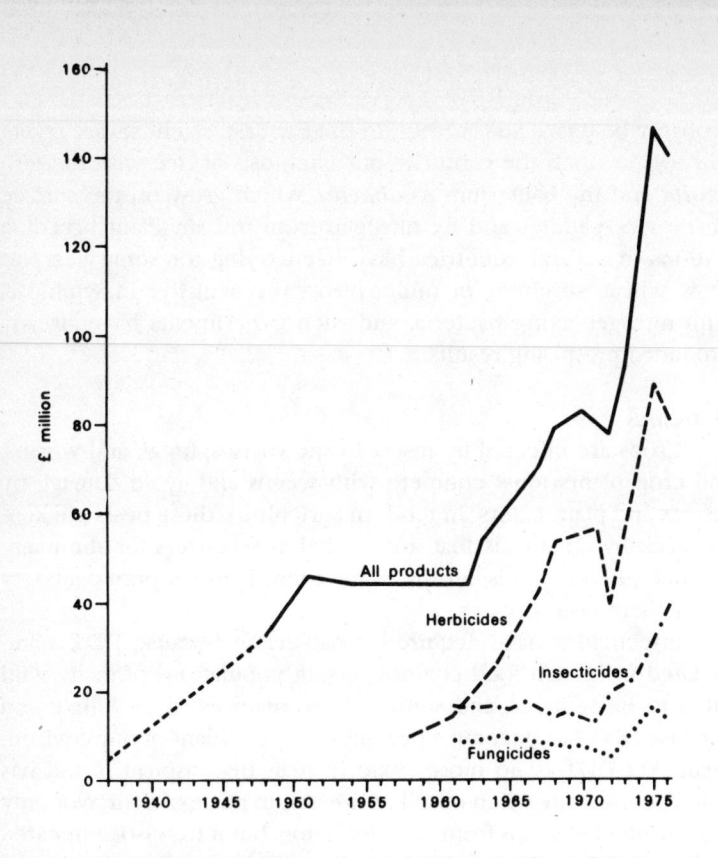

7. Consumption of pesticides and weed killers in Great Britain, 1937–1976. SOURCE: *Seventh Report of the Royal Commission for Environmental Pollution, Agriculture, and the Environment* (Her Majesty's Stationery Office, Cmd. No. 7644, 1980).

the thinning of the eggshells of wild birds. DDT has probably saved more human lives than all the antibiotics combined (figure 9).[18] There has also been anxiety that DDT in the ocean might become concentrated in the phytoplankton on which the food chain depends, but in fact its concentration has never reached one-thousandth of the danger level, and it has been falling in recent years.[19] DDT remains the cheapest pesticide for use against malaria, but DDT and other organochlorides are not now officially recommended as agricultural pesticides and are no longer manufactured in Britain, where its use as an agricultural pesticide had reduced the population of wild birds, especially birds of prey such as the sparrow hawk and the peregrine falcon. Their populations have now recovered to pre-DDT levels. Farmers are told to use organophosphates and carbamates, which do not enter our diet or accumulate in the food chain because they are quickly decomposed. But some of these substances are very toxic and must be handled with great care by those who apply them. Even in Britain the official recommendations do not have the force of law, and consequently some farmers still use imported organochlorides, regardless of the damage they do to wildlife, because they are cheaper than the recommended pesticides.[20]

Herbicides have acquired a bad name since American forces used 2,4,5-trichlorophenoxyacetic acid (2,4,5-T) as a defoliant in Vietnam at far higher concentrations than are used in agriculture. These sprays often contained the toxic impurity dioxin, notorious also for its release at Seveso.[21] Most herbicides now used in agriculture do not accumulate because they are broken down by soil bacteria, and they present no danger to those who eat the agricul-

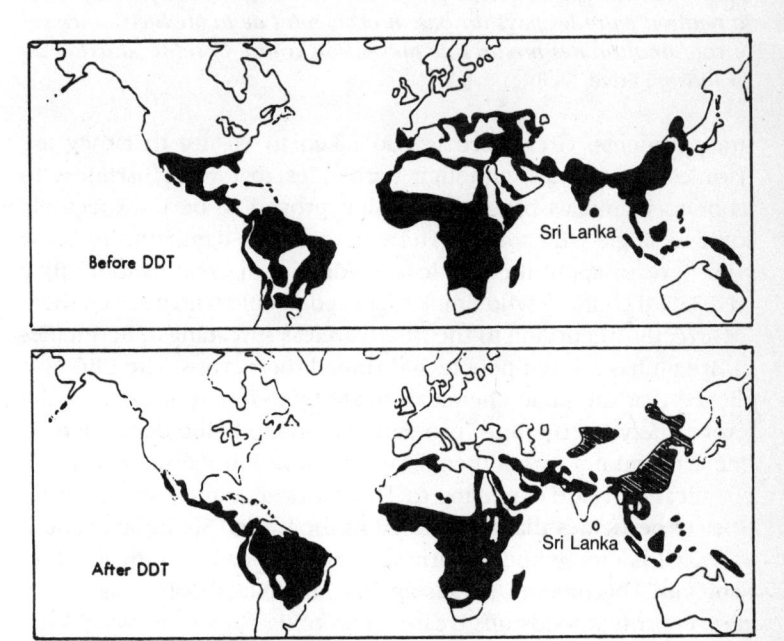

Black shaded areas = regions where malaria is endemic
Partially shaded areas = regions where information about malaria incidence is unavailable

8. Elimination of malaria by spraying with DDT. SOURCE: M. W. Service, "Control of Malaria," in *Ecological Effects of Pesticides,* eds. F. R. Perring and Kenneth Mellanby (New York: Academic Press, 1977).

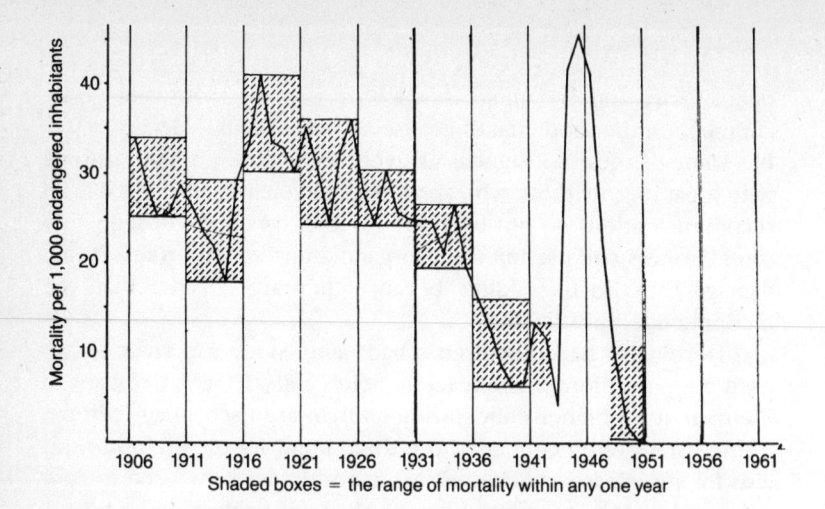

9. Mortality from malaria in Italy per 1,000 endangered inhabitants, 1906–1961. SOURCE: *Rapport rédigé par la délégation italienne participant à la réunion entre les pays du bassin occidental de la Méditerrannée sur la coordination des mesures de prévention contre la reintroduction du paludism* (Erice, Sicily, 1979).

tural products. Great care is also taken to ensure that they are harmless to animals, although herbicides that were harmless to laboratory animals have on occasion proved to be unexpectedly toxic to single wild species. Herbicides can be dangerous to those who have to apply them unless handled with great caution; they have killed children who drank them and people who handled them incorrectly. According to the BBC, careless spreading of herbicides in Britain has injured people and ruined their crops. The BBC also alleged that the same chemicals made by different firms carry different safety instructions: protective clothing being demanded by one firm and not by another. Two American journalists report that chemical firms are exporting to Third World countries large quantities of pesticides that are banned in the United States on account of their toxicity, without warnings understandable to the local inhabitants. This practice is causing many often fatal poisonings. They also report that foodstuffs treated with toxic chemicals were being imported into Western countries without their contamination being tested. I have not been able to check these accusations with independent sources; some, though not all of them, seem well documented.[22]

A British medical research worker investigated the changes in mortality in a province of the Philippines during a period when there had been a large rise in the use of insecticides. He discovered a great increase in the mortality of male farm workers who sprayed the insecticides without protective clothing from packs carried on their backs, while the mortality of women and children in the countryside and that of male workers in neighboring towns dropped during the same period.[23] In Sri Lanka, 13,000 people are admitted to hospitals for pesticide poisoning each year, and about 1,000 die from such poisoning. According to reports in English newspapers, speakers at a meeting of the FAO estimated that 40,000 farm workers were killed by pesticides in the Third World each year, and that injuries to health were sustained by a multiple of that number. Farmers are poisoned because they cannot read the safety instructions, cannot understand them, or lack the means of implementing them, or because the instructions are inadequate. Here the conflict between benefits and risks reaches tragic proportions: the hugely swollen populations of the Third World cannot be fed without intensive agriculture, for which pest control is essential; people are too poor and ignorant to protect themselves when they spray pesticides. Their poverty and ignorance is a consequence of the fast rise in population. That rise in turn is a result of Western medicine and public health, whose introduction we regard as our most beneficial contribution to the rest of the world.

The FAO, in collaboration with the international trade association of manufacturers of agrochemicals, is trying to halt this poisoning by laying down an "International Code of Conduct on the Distribution and Use of Pesticides." This code lists a series of rules for governments and industry: for example, "to introduce and implement laws, regulations and controls . . . for the registration of a product before it is allowed to be sold." It calls upon industry "to follow the product up to the end-user in order to find out whether there is a need to make changes in the type of formulation, the package, the labelling or the distribution; . . . not to sell products in tropical countries which require uncomfortable protective clothing; . . . to disseminate educational materials to pesticide users; . . . to train persons who sell pesticides in giving advice on safe and efficient use." Shell Chemicals already supplies such advice in the form of pictograms for people who are illiterate. It is to be hoped that adherence to the code will make it possible to maintain high yields without sacrificing human health.

The chemical industry is developing new generations of pesticides, which should be less toxic, less polluting, and active at smaller concentrations than the present ones, and is also improving the methods of application. For example, scientists at the Rothampstead Experimental Station in England have developed the pyrethroid insecticides, which are analogues of the natural agent pyrethrum extracted from an African daisy and are nontoxic to mammals. They are now being used worldwide, and the search for other insecticidal plant products is growing. The British firm Imperial Chemical Industries has developed an electrostatic sprayer that achieves effective pest control at less than 1 liter instead of the usual 1,000 to 2,000 liters per hectare.[24] This machine has proved successful in several Asian and African countries and minimizes the noxious side effects of chemical pest control, even though its use could not prevent the growth of pesticide-resistant mutants. In general, it is desirable to keep application of pesticides to a minimum. In Switzerland, chemical control of wheat pests is therefore recommended only when they cause a loss of more than fifty dollars' worth of harvest per hectare of land. This criterion is also logical because otherwise the application of pesticide would cost more than the likely loss.

In Britain the greatest danger to wildlife has come from financial incentives that induced farmers to grow food on more of the land on which wild animals lived and to sow meadows on which wildflowers grew with rye grass for cattle. Hedges and trees where birds nested have been cut down; stubble, and with it all the creatures that lived in the field, has been burned after the harvest. Worldwide, all wild species are threatened by the insatiable hunger for land of the ever-growing multitudes of people.

The real danger to man lies not in the pollution of the environment by pesticides, but in the selection and proliferation of pesticide-resistant mutants of insects and fungi. Research workers are trying to overcome this problem by synthesizing new pesticides, but there may be limits. In 1956 the synthesis of 1,800 chemicals led to 1 commercial pesticide; in 1967 the ratio was 5,000 to 1; in 1976 it grew to 10,000 to 1. When a promising compound has at last been found, it must be subjected to elaborate tests of its possible adverse effects over a period of years. These include toxicology and feeding tests for animals, birds, fish, bees, and microorganisms in soil and water; large-scale farm trials; and trials at government experimental stations. The collected data are then submitted to the British government's Pesticides Safety Precautions Scheme for clearance and registration. As a result, an outlay of $30 million and six to seven years are now needed in Britain to produce a marketable compound. Such large investments can be recovered only for mass products. They discourage the development of a variety of selective insecticides, which can be applied in low concentrations to kill specific insect pests and leave their predators and other useful insects alive. Public insistence on absolute safety is therefore threatening to defeat its own object.

What alternatives are there? Biological methods of pest control, such as the breeding and release of sterile male insects for mating with the females, may soon help to keep some insect pests at bay. Another biological method consists of spreading viruses that attack insect pests. The most promising are the baculoviruses that infect certain insect larvae and grow in them to form microscopically visible polyhedral inclusion bodies that fill and kill the larvae. Baculoviruses attack only the larvae of particular pests, not those of bees and other useful insects, and they are harmless to animals and people.

Biological methods often require detailed interdisciplinary studies of the behavior, physiology, and biochemistry of the relevant insects under natural conditions. Elizabeth Bernays, a pioneer in the ecology of insect pests, studied the destruction of the cassava harvest by grasshoppers in Nigeria. She found that the plants defended themselves by producing cyanic acid, which surprisingly did not kill the grasshoppers, but because the grasshoppers appeared to dislike its bitter taste so much, they starved rather than ate a healthy cassava plant. As long as the fields were well irrigated, the grasshoppers preferred eating the weeds that grew between the cassava plants. However, when the soil was dry, they bit the cassava plants, and a hundred bites suffice to make a plant wilt. Wilted plants yielded little cyanic acid and were soon eaten. Bernays therefore advised the farmers to keep their fields well irrigated to make the weeds grow, but the poorer farmers could not afford to do this. Bernays then discovered that the grasshoppers laid their eggs in large masses under the surface of the soil. When she dug the eggs out, they were killed by the sun. Even the poorest farmers could afford to do this.[25]

Plants fight their own chemical war against parasites: by genetic mutation they evolve new toxins, to which the insects adapt by their own mutations, just as the chemical industry develops new pesticides, against which the insects evolve new resistant mutants.

Plant breeders exploit the ability of plants to produce insect poisons in order to select and breed pest-resistant varieties of crops; they have been successful for some pests but not for all. Another way of preventing the growth of pesticide-resistant mutants is by using pesticides at concentrations low enough to allow some fraction of wild-type organisms to survive. Since the wild type is generally fitter than the pesticide-resistant mutant, it will outgrow the mutant in the absence of further pesticides, so after an interval of time the same pesticide will again be effective in reducing the pest's numbers.

If all agricultural sprays had to be abandoned in Britain today, cereal crops would be reduced by 24 percent in the first year, mainly by pest attack, and by 45 percent in the third year, mainly by weeds,[26] but these losses might be reduced by the planting of pest-resistant varieties and better crop management. Both the Irish potato famine of the last century and the great Bengal famine of the 1930s were caused by fungi attacking crops. If we reverted to organic farming without fungicides, we would not be able to prevent a repetition of these disasters. Fungal infestations of certain crops after harvesting produce the most deadly carcinogen known, aflatoxin, which causes cancer of the liver in many tropical countries, where grain cannot be stored dry.

In a discussion of pesticides at a London symposium on Better Crops for Food, the Dutch agriculturalist J. C. Zadoks said,

The world's major food crops are annual plants. With annuals, breeding for resistance is relatively easy, be it against fungi, insects, nematodes or viruses. Resistance is and must remain the first line of defence against all harmful agents. However, resistance based on transferable genetic characteristics does not always suffice, for a variety of reasons. New strains of well-known harmful agents appear regularly, destroying the effectiveness of laboriously obtained resistance. Once in a while, new harmful agents appear. Formerly, cultural methods were the most important, though often implicit, method of crop protection. The cultural methods of intensive agriculture, however, are counter-productive here. Irrigation causes lush growth, and the split application of nitrogen causes leaves to last longer, producing a paradise for harmful agents. Biological control has not yet developed sufficiently to safeguard the world's food supply, although there have been incidental successes which can be regarded as promises for the future. Crop protection chemicals are with us to stay.[27]

While millions of people are starving, all over the world, nearly half of agricultural produce (in some countries much more) is believed to be destroyed by pests, much of it in storage.[28] We need more effective pest control, which could often be achieved with smaller quantities of selective pesticides. I have come to realize that my original views on the dangers of pollution by agrochemicals were formed by people who publicized their adverse effects and concealed their benefits. Provided they are manufactured and handled with care and applied discriminately, chemical pesticides present little danger to man or animal, and they have multiplied the crops on which our lives depend. Concern should arise from the careless practices used in the manufacture of pesticides and herbicides like those that caused the disasters at Bhopal and Seveso, as well as the discharge of poisons into the Rhine at Basel; from the export of dangerous pesticides to countries where people do not know how to handle them; from the huge amount of energy required by modern agriculture; and from the development of pesticide resistance. In the United States 10 calories of energy are used to produce and distribute 1 calorie of food. In rich countries that energy is merely a small fraction of the total consumed, but in poorer countries any shortage of energy that drives up its already high cost would exacerbate the shortage of food.

BIOTECHNOLOGY AND GENETIC ENGINEERING IN AGRICULTURE

Will genetic engineering lead to better methods of pest control? For the moment, there are only hopeful beginnings, but the chemical industry is already investing huge sums of money in them. The work rests mainly on two basic discoveries: the cloning of plant cells and the transfer of new genes to plants by a tumor-producing organism (*Agrobacterium tumefaciens*).[29]

Plant cells were first cloned in 1958 by the English biologist Frederick C. Steward at Cornell University and independently by the German biologist Jacob Reinert, who discovered a method of growing complete carrot plants from cultures of single carrot cells. The potential for plant breeding was soon recognized and led to the development of clones of pollen cells or somaclones (so called because they are derived from somatic cells, cells from roots or leaves, as opposed to pollen or seeds). Such clones were developed in many food plants, including wheat, maize, rice, sugarcane, potatoes, tobacco, strawberries, and pineapples, but also in horticultural crops, such as orchids, chrysanthemums, and petunias. Cells in culture are subject to mutations, which may give rise to useful genetic variants. Growth of a clone under selective pressure, for example, in the presence of the toxin of a fungal pest or a weed killer, may produce heritably resistant variants faster than the classic method of making hybrids from crosses. However, crosses have the advantage of uniting the desirable properties of two genetically different variants. In 1960 the English biologist E. C. Cocking opened the way to making crosses in cell cultures. He used an enzyme to digest away the cellulose walls from the cells of the root tips of a tomato plant; the digestion left naked protoplasts that could be fused to protoplasts of other variants. From such fused cells useful hybrids could be grown. In Hawaii and Fiji somaclones have been used to grow sugarcane that is resistant to the Fiji virus and a local fungus (*Sclerospora sacchari*). In Australia a variant sugarcane resistant to eyespot disease (*Helminthosporium sacchari*) was produced. It has also been possible to grow several hundred somaclones from a single wheat embryo and to isolate many variants in only two to three months, much less than the time needed to produce variants from seed.

New genes can be introduced by agrobacteria into either plant embryos or single cells. Such gene technology was employed to make tobacco plants poisonous to butterfly larvae. *Bacillus thuringiensis kurstaki* was known to make a protein toxic to butterfly larvae but harmless to animals and man. The gene for this protein was cloned in coli bacteria and transferred to tobacco plants by agrobacteria; these plants now produce the toxin in their leaves from generation to generation. The chemical firm Monsanto has gained official permission to plant such tobacco in trials in open fields.[30]

Another group has tried to "vaccinate" tobacco plants against the tobacco mosaic virus. They cloned the gene for the viral coat protein, transferred it to a piece of DNA of an *Agrobacterium* from which the tumor-producing genes had been cut away, and then introduced that gene into tobacco plants. These genetically transformed plants made the viral coat protein without the virus. When seedlings carrying the coat protein gene were inoculated with small doses of live tobacco mosaic virus, they rarely developed symptoms; when larger doses were used, transformed plants took longer than untransformed ones to develop symptoms.[31] In the past, people had tried to immunize plants to viral diseases by inoculation with harmless virus variants, but these inoculations ran the danger of mutations turning the harmless viruses into virulent ones or of making a virus that protected one plant but caused disease in another. The new methods avoid these dangers, since they omit the genes for viral replication.

Yet another group has bred petunias that are resistant to the weed killer Glyphosate, an effective and much used herbicide that also kills most useful plants. Its action is based on inhibition of an essential enzyme. Genetic engineers devised a variant petunia that contains twenty copies of the gene coding for that enzyme and therefore produces it in sufficient abundance to be resistant to the herbicide.[32]

Agrobacterium does not infect cereals, which makes the usual method of gene transfer inapplicable to them. Instead, a German group injected a new gene into rye directly. To test their method, they injected many copies of a piece of DNA containing the gene for resistance to the antibiotic kanamycin into the lateral buds of rye seedlings. After fertilization by pollen of similarly inoculated plants, the seeds were harvested and sown on soil containing kanamycin. Seven out of 3,000 budding seeds grew seedlings that stayed green, and two of these contained the enzyme that confers resistance to kanamycin. Their seeds would pass that resistance on to their descendants. This experiment has proved that it is possible to introduce new genetic information into the sex cells of cereals and to harvest seeds from which normal plants expressing the new gene can be grown. Such techniques may facilitate the introduction of resistance to viruses and fungi[33] and reduce the need for pesticides.

Genetic variation of somaclones, breeding of mutant plants, and transfer of genes into plants have already led to the production of a few new varieties that can be propagated from generation to generation; these methods can be expected to become increasingly useful to plant breeders. Nevertheless, the greatest successes in plant breeding so far have been achieved with classic methods of crossing and selection.[34] For example, from crosses of bread wheat with a wild grass (*Aegilops ventricosa*) made in France, scientists at the Plant Breeding Institute in Cambridge, England, have produced a new variety of wheat that is resistant to eyespot, a frequent fungal disease.[35] They are also developing salt-resistant lines of wheat using wide hybridization to transfer genes from distantly related grasses. So far as I know, all new varieties of cereals that have contributed to the Green Revolution have been bred by classic methods, but twelve to fifteen years usually elapse from the first growth of a new cross to its cultivation by farmers. Biotechnologists hope to shorten that period, especially because new mutant pest populations adapted to a pest-resistant variety often arise in less than four years.

Applications of biotechnology to animal husbandry have so far been restricted to artificial egg and embryo manipulations, insemination, and treatments with bioengineered hormones and vaccines, for instance, against foot-and-mouth disease. Future possibilities can be gauged from present experiments with mice.

The American biologist Beatrice Mintz has devised an ingenious method for introducing new genes into mouse embryos.[36] Cells of certain mouse cancers propagated in cultures and implanted into fully grown mice give rise to malignant tumors. Mintz injected a single such cell into a mouse embryo consisting of only a few cells. That embryo developed normally, without malignant growths, and the mouse was a hybrid made up of descendants of the original embryonic cells and of the injected cancer cell. When the embryonic cells came from a pair of white mice and the cancer cell from a black one, the young mice were striped black and white. Over 70 percent of the young contained both kinds of cell in all their tissues. When such hybrid mice were mated for two generations with normal inbred mice, genetic segregation led to purebred descendants of the cancer cells.

Cultures of cancer cells lend themselves to genetic engineering; chemical and genetic methods can ascertain whether a newly introduced gene is inserted correctly into the desired chromosome. Mintz's method therefore opens the way to the breeding of animals from a clone of genetically transformed cells.

R. L. Brinster and his collaborators in the United States have developed another method for transferring genes into mice. They injected a suspension of DNA fragments into fertilized mouse eggs. The fragments contained the desired gene coupled to the gene for an enzyme that is activated by zinc salts. If zinc is later added to the food of the young mice, the proteins coded for by the introduced genes are made in the mice's tissues. Introduction of the gene for the human growth hormone followed by administration of zinc salts raised the hormone concentration in such genetically transformed young mice up to eight hundred times compared with the concentration in their normal siblings; some of the mice also weighed nearly twice as much as their normal siblings. One of these animals transferred the newly introduced genes to half its descendants, from which further generations inherited the genes.[37]

Such methods of genetic engineering will probably be applied to the breeding of farm animals. Will they also be used for gene therapy of humans? To me, this seems out of the question for the following reason. Brinster and his colleagues injected about 600 copies of the new genes into each of 170 eggs, which they then implanted into the uteri of 170 mice. From these only 21 young mice developed, and only 6 of these were larger than their untreated siblings. In Mintz's experiments only 390 of 1,258 embryos into which the cancer cells had been implanted survived; 78 of those that lived to full term died young of a virus epidemic; of the 312 survivors of that epidemic, only 41 showed the dark color derived from the cancer cells. It would be criminal to try methods on humans in which not more than a fraction of the fertilized eggs develop normally and the desired gene therapy works in only a fraction of that fraction.

SCIENCE AND HEALTH

CAUSES OF DEATH

People often look back nostalgically to the good old days without hustle, noise, and smog, but they forget the much greater risk of early death that weighed on our forebearers and found its expression in poetry and religious thought. In 1693 the English astronomer Edmond Halley published the results of his study of life expectancies in the German city of Breslau, where good records of births and deaths were kept, with a view to calculating the price of annuities. Figure 10 plots his results. Of every 100 children born, only 51 were alive at the age of ten; only 43 reached the age of thirty; only 28 reached the age of fifty; and only 11 reached the age of seventy.[38] Life expectancy was not much better in England in 1867, when half of all deaths occurred among children below the age of fourteen.[39] Table 2 lists some of the many artists, musicians, and writers of the past who were killed by infectious diseases in the prime of life. Masaccio, the pioneer of Renaissance painting, died of the plague at twenty-seven; Mozart died at thirty-five of an undiagnosed

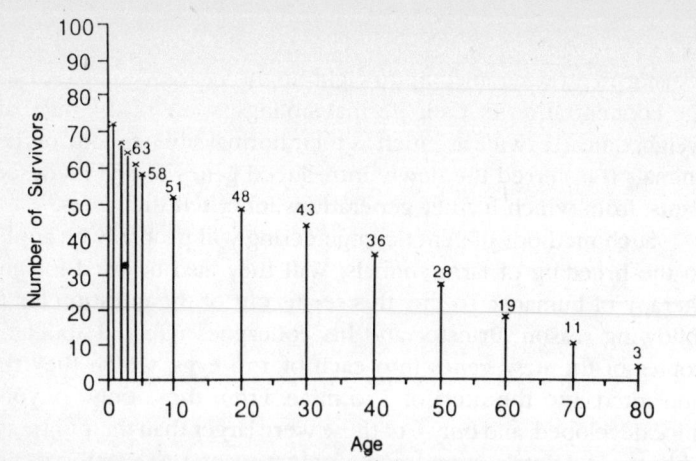

10. Halley's calculation of life expectancies in Breslau, 1690. SOURCE: *Philosophical Transactions of the Royal Society* 17 (1693): 596.

TABLE 2

"THE GOOD OLD TIMES": CAUSE OF AND
AGE AT DEATH OF GREAT MEN AND WOMEN

	PROFESSION	YEAR OF BIRTH	AGE AT DEATH	CAUSE OF DEATH
Masaccio	Painter	1401	27	Plague
Giorgione	Painter	1477	33	Plague
Raphael	Painter	1483	37	Sudden fever
Wolfgang Amadeus Mozart	Composer	1756	35	Fever
John Keats	Poet	1795	36	Tuberculosis
Heinrich Heine	Poet	1797	59	Tuberculosis
Franz Schubert	Composer	1797	31	Typhoid fever
Frédéric Chopin	Composer	1810	39	Tuberculosis
Robert Schumann	Composer	1810	39	Syphilis
Emily Brontë	Writer	1818	22	Tuberculosis
Anne Brontë	Writer	1820	29	Tuberculosis
Charles Baudelaire	Writer	1821	46	Syphilis
Friedrich Nietzsche	Poet, philosopher	1844	56	Syphilis
Paul Gauguin	Painter	1848	55	Syphilis
Georges Seurat	Painter	1859	31	Laryngitis
Hugo Wolf	Composer	1860	43	Syphilis
D. H. Lawrence	Writer	1885	45	Tuberculosis
George Orwell	Writer	1903	47	Tuberculosis

infection, just after he had finished *The Magic Flute* (there is no evidence of his having been poisoned by Salieri or anyone else); Schubert died of typhoid at thirty-one, when his music had reached the depth and perfection of Beethoven's. It is tragic to think how many great works have been lost because of diseases that can now be prevented or cured. Of course not all great men died young: Leonardo reached the age of sixty-seven, Titian eighty-seven, Galileo seventy-eight, and Newton seventy-four, but their chance of an early death was much greater than ours.

In developed countries public health has improved beyond the most sanguine expectations of fifty years ago. Who would have thought that tuberculosis, smallpox, and many other infectious diseases would practically disappear (figure 11); that the average life expectancy even in Western Europe would rise by another ten years. In Britain it is now seventy years for men and seventy-six years for women, but this is an average of a distribution that varies with social class. Life expectancy is ten years shorter among unskilled workers than among the managerial and professional class.[40] In the United States, life expectancy has risen steadily throughout this century, and it continues to rise by about three years in every ten years. Despite widespread poverty, life expectancy is still rising steeply all over the world, except in the Communist states of Eastern Europe and the Soviet Union, and it is higher in India today than it was in any European country at any time in the nineteenth century (figure 12).[41] The most striking rise has occurred in Japan. There it is accompanied by an increase in the average height of young Japanese, which now equals that of Europeans, and also in the average IQ of Japanese schoolchildren, which is now supposed to exceed that of their European and American counterparts (figure 13).[42] These are the most convincing testimonies for the benefits of Western diet and public health.

It is often said that life expectancy in Europe rose and mortality from infectious diseases fell long before the introduction of antibiotics, and therefore had little to do with research, but the example of tuberculosis shows that this is only partly true. Mortality from tuberculosis has diminished since the beginning of this century thanks to better hygiene and improved living standards, but from the moment antibiotics were introduced, the fall became much steeper, and now mortality from tuberculosis in developed countries is close to zero (figure 14). Improved hygiene owes much to

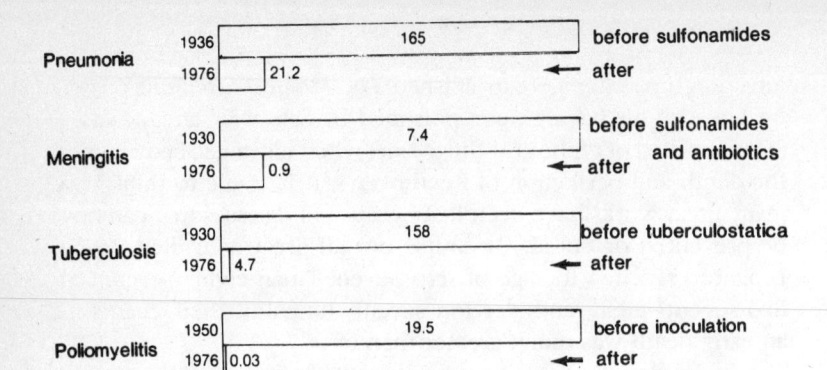

11. Annual mortality per 100,000 Germans from four infectious diseases, 1930–1976. SOURCE: *Arzneimittel-forschung in Deutschland* (Pharma, Bundesverband der Pharmazeutischen Industrie, Karlstrasse 21, 6000 Frankfurt, 1979–1980)

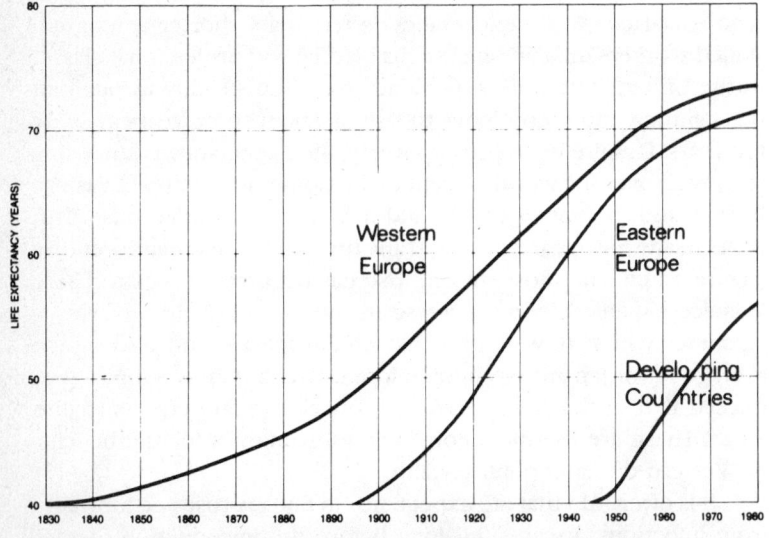

12. Average life expectancy in Western and Eastern Europe and in developing countries, 1830–1980. Hygiene and antibiotics have raised the average life expectancy in developing countries by as much in fifteen years as the combination of hygiene, improved nutrition, and old-fashioned medicine did in Europe in eighty-five years. Even so, life expectancy in developing countries still lags behind that in Europe by fifteen to twenty years, mainly because of higher infant mortality. SOURCE: D. R. Gwadkin and S. K. Brandel, "Life Expectancy and Population Growth in the Third World," *Scientific American* 246 (May 1982): 33.

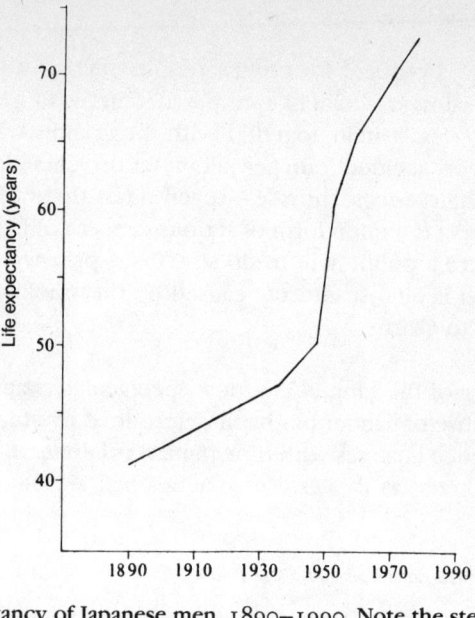

13. Life expectancy of Japanese men, 1890–1990. Note the steep increase as a result of American influence following the Second World War. SOURCE: A. M. Anderson, "The Great Japanese IQ Increase," *Nature* [London] 297 (1982): 181.

Pasteur's, Koch's, Semmelweis's and others' research on the bacterial origins of many diseases.

Can medicine advance further? In the United States today, half of all deaths are caused by arterial disease and one-third by cancer, but these figures are misleading because they do not tell us at what ages people die from these causes. John Cairns has obtained a measure of the untimeliness of death by calculating the total loss of working life span for those between the ages of twenty and sixty-five from different causes (table 3). When individual causes of death are analyzed in this way, infant mortality still tops the list, followed by cancer, heart disease, and automobile accidents. Thanks to antibiotics, infectious diseases are responsible for no more than 2 percent of untimely deaths, fewer than those caused by homicide. In America suicides account for 3 percent of the years of working lives lost; in Britain they account for 1 percent of all deaths, and for as much as 12 percent of deaths between the ages of twenty-five and twenty-nine. This tragic loss of young lives is a great challenge to psychiatrists and social workers.[43] In the United States, automobile accidents cause 11 percent of all untimely deaths, compared with 13 percent for cancer, despite the fact that the United States has a lower accident rate per kilometer driven than many other countries. Britain, together with the Scandinavian countries, has the lowest accident rate per kilometer driven; Poland and Spain have the highest accident rate—seven times that of Britain!

The most common form of serious cancer could be prevented if there were a public will to do so. This is primary cancer of the lung, which is almost entirely caused by the smoking of tobacco. According to Cairns,

> Cancer of the lung is the most spectacular example in which the cause of cancer has been determined by studying the way incidence changes with time [figure 15]. Indeed, in retrospect, it is almost as if Western societies had set out to conduct a

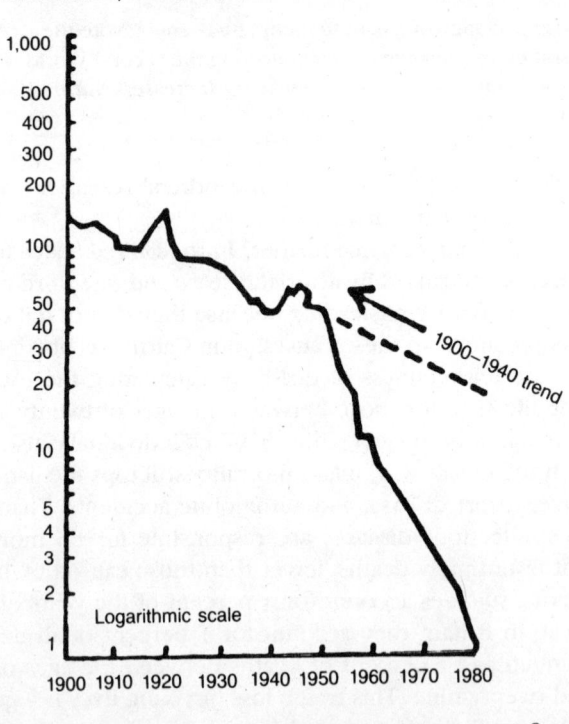

Annual deaths per 100,000 population

14. Mortality from tuberculosis in England and Wales, 1900–1980. SOURCE: British Department of Health and Social Security.

TABLE 3

LOSS OF WORK YEARS IN THE UNITED STATES
FROM VARIOUS CAUSES, 1968

	WORK YEARS LOST*	% OF TOTAL
Accidents and violence:		
Automobile accidents	1,533,102	11
Other accidents	1,262,415	9
Homicide	397,668	3
Suicide	389,733	3
SUB TOTAL	3,582,918	26
Vascular diseases:		
Heart disease	1,610,142	12
Cerebral vascular diseases	431,973	3
Other	578,801	4
SUB TOTAL	2,620,916	19
Infant mortality	1,970,489	14
Cancer	1,744,189	13
Respiratory diseases	968,064	7
Congenital diseases	674,465	5
Infectious diseases	291,185	2
TOTAL	11,852,226	86
All causes	13,687,716	100

*Working life is considered to extend for forty-five years, from age twenty to age sixty-five. Deaths occurring before the age of twenty each contribute forty-five lost years to the total, those occurring between twenty and sixty-five contribute appropriately fewer, and deaths after sixty-five do not count.

SOURCE: John Cairns, *Cancer, Science, and Society* (San Francisco: W. H. Freeman, 1978).

---

vast and fairly well controlled experiment in carcinogenesis bringing about several million deaths and using their own people as experimental animals.[44]

Cancer of the lung is not the only disease related to smoking. In Britain deaths from bronchitis and pneumonia are as frequent as those caused by cancer, and a large proportion of these deaths and of those from cardiovascular diseases are also attributable to smoking. Deaths from cirrhosis of the liver related to alcoholism are

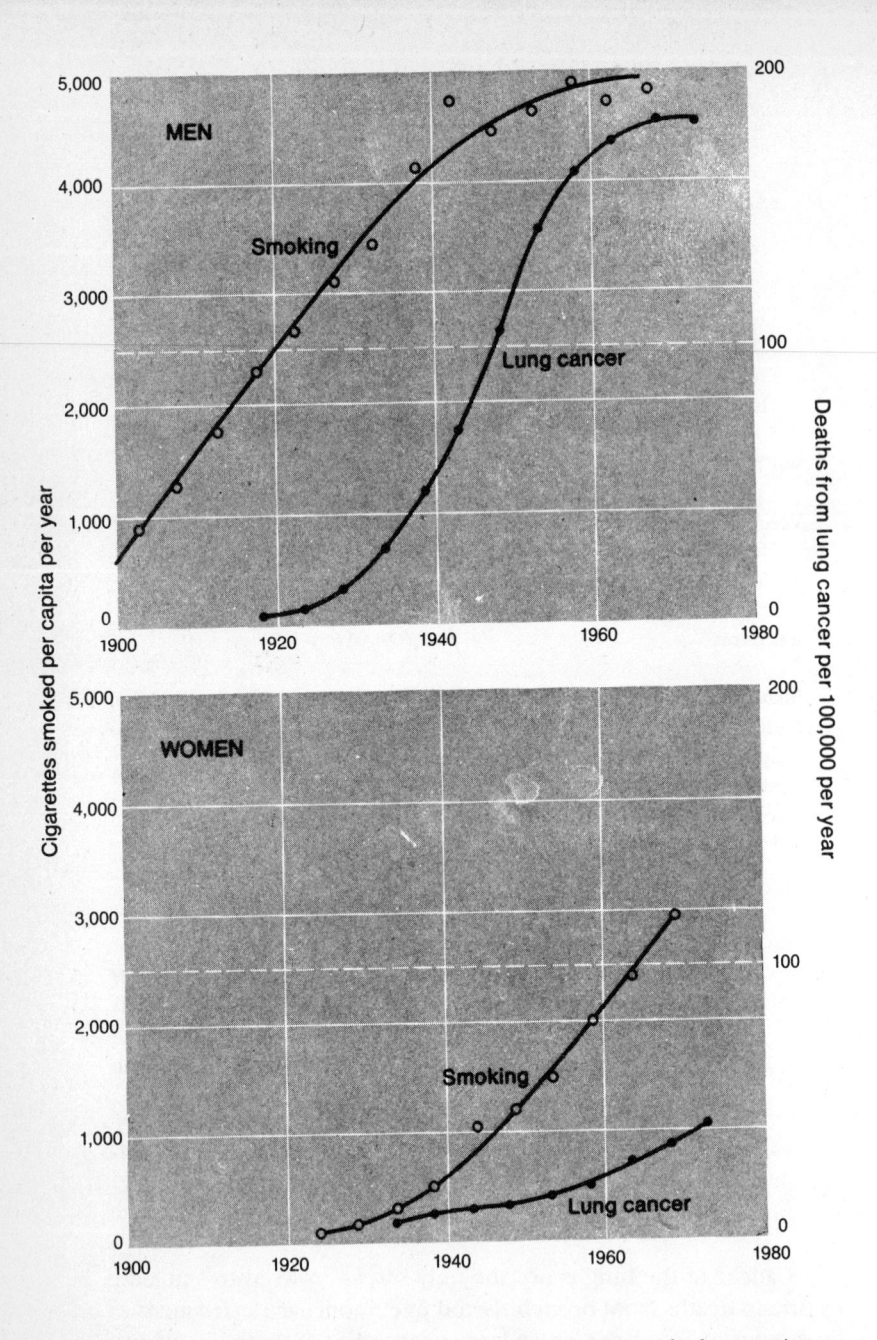

15. Mortality from lung cancer compared with cigarettes smoked per capita in the United States, 1900–1980. SOURCE: John Cairns, *Cancer, Science, and Society* (San Francisco: W. H. Freeman, 1978).

	0	10,000	20,000	30,000	40,000	50,000	60,000	70,000

Mouth, esophagus, pharynx, or larynx

Lung

Stomach

Intestines (including rectum)

Liver, gallbladder, and bile ducts

Pancreas

Bone

Melanoma

Breast

Bladder

Kidney

Cervix uteri

Endometrium and other uterus

Ovary

Brain and nervous

Leukemia

Hodgkin's disease

Other reticuloendothelial

Other and unspecified solid (except skin)

Key ▬ (Second National Cancer Survey) 1947–1948

▬ (Third National Cancer Survey) 1969–1971

Female, all ages, registered incidence

16. Incidence of cancer per 100 million women of all ages in the United States, 1947–1948 to 1969–1971. Note the rise in cancer of the lung caused by increased smoking of cigarettes and the drop in cancer of the stomach and of the cervix, for reasons unknown. SOURCE: Richard Doll and Richard Peto, *The Causes of Cancer* (Oxford: Oxford University Press, 1981).

rising, especially in Scotland.[45] The British Department of Health spends £1.5 million annually on its antismoking campaign, while the tobacco firms spend £80 million on advertising cigarettes In Britain, antivivisectionists demonstrate outside medical laboratories against the use of animals for cancer research, but I have never heard of anyone demonstrating outside automobile factories against producing cars that crush people on impact, or outside cigarette factories to stop the appalling epidemic of cancer of the lung that kills people in many countries (see "Department of Defense," pp. 226–27).

Cancer is widely believed to be caused also by chemical food

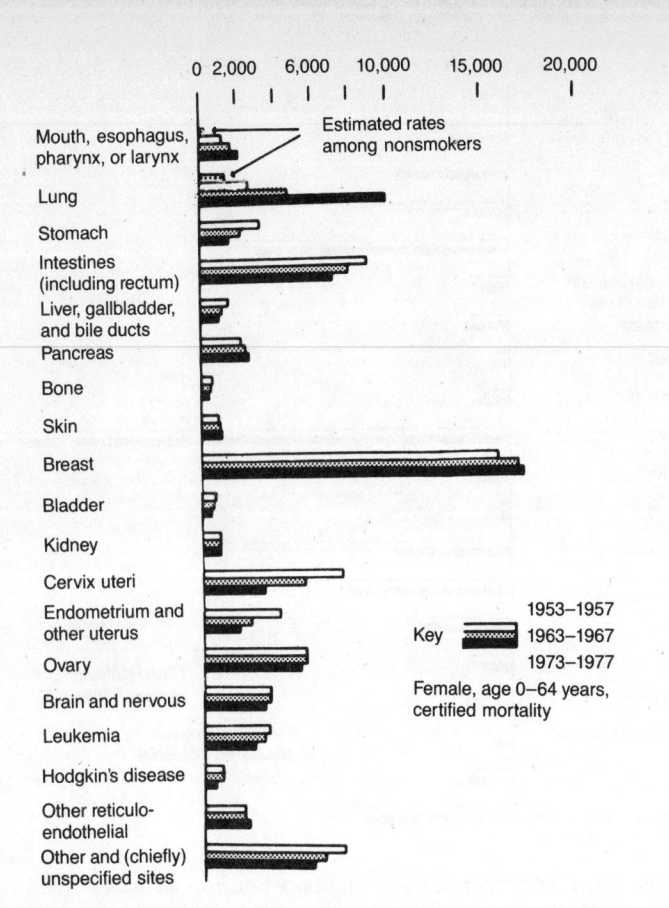

17. Mortality from cancer per 100 million American women between the ages of zero and sixty-four years, 1953–1957 to 1973–1977. Note the large rise in cancer of the lung. SOURCE: Richard Doll and Richard Peto, *The Causes of Cancer* (Oxford: Oxford University Press, 1981).

additives and other industrial products, but the leading English epidemiologists Richard Doll and Richard Peto found no indications of this being true, except for certain industrial workers. Figures 16 and 17 show that in the last twenty-five years cancer of the respiratory tract is the only form of cancer whose incidence has increased markedly, and this has been proved to be attributable exclusively to the increased smoking of cigarettes. Mortality from breast cancer has increased slightly, possibly because women are older when they have their first child, but the incidence of and mortality from other cancers have remained the same or have fallen, whereas, if the environmentalists' accusations were right, increased use of industrial products ought to have made at least some of them rise.[46]

The real causes of cancer probably lie elsewhere. For example, the incidence of melanomas of the skin has risen sharply, especially in California and other sunny regions, apparently as a result of excessive sunbathing. Doll and Peto found large geographic variations in the incidence of different cancers. Skin cancer is two hundred times more common in Queensland, Australia, where white-skinned people do too much sunbathing, than in Bombay, India, where people's brown skin protects them from the sun. In Mozambique, primary cancer of the liver is a hundred times more common than in England, perhaps because in Mozambique infection with hepatitis B is common. Cancer of the prostate is forty times more common among American blacks than among native Japanese, for reasons unknown. Breast cancer is seven times more common among women in British Columbia than among non-Jewish women in Israel. Bladder cancer is six times more common in Connecticut than in Japan, for reasons unknown. These variations led Doll and Peto to conclude that many cancers have causes that are linked to people's traditional life-styles, and that much cancer could be prevented by the discovery of these causes. For example, the Japanese oncologist Takashi Sugimura has detected carcinogens in meat or fish that has been roasted in charcoal fires, and he regards these as a major source of intestinal cancer.[47] According to him barbecues are to be avoided.

A massive research effort is being made to find the causes of various other types of cancer and to prolong the lives of cancer patients. This is as it should be, but Cairns's figures make me wonder why a much larger effort is not also being made to reduce the comparable number of untimely deaths and the much larger incidence of disability caused by road accidents. The problems involved in such an effort would be much easier and cheaper to solve than those of cancer. In fact, many of the solutions probably exist, but the political will to apply them is lacking. For instance, until recently the British Parliament believed that compulsory wearing of seat belts, which is known to reduce the severity of injuries in car accidents, would be an infringement of personal freedom. In Italy most cars are not even equipped with seat belts. The speed limit of 55 miles per hour, which is enforced in many American states, is known to reduce accidents drastically. In the year after its introduction there were 9,343 fewer traffic accidents, a reduction of 17 percent; there were estimated to have been 90,000 fewer head injuries leading to epilepsy, and 60 to 70 percent fewer cases of paralysis resulting from back injuries. These figures were supplied to me by the British Ministry of Transport, but neither Britain nor any other Western European country has followed the American example, and speed limits were raised on some American roads in 1987. Many road accidents are related to alcohol (30 percent in Britain) and could be prevented by stricter control of driving under its influence. Unlike animal rights, the killing and crippling of people in traffic accidents is not a political issue; speed limits and alcohol checks are resented as infringements of personal liberty. (Italy has now introduced speed limits of 70 miles per hour on freeways and 55 miles per hour on other roads. Britain has the same limits, but they are not enforced any longer.)

DRUGS AND VACCINES

Cancer therapy and the treatment of many other disorders depend on the development of new drugs. Chemotherapy with drugs that inhibit cell division has made it possible to save the lives of many leukemic children. In America in the early 1950s about 1,900 children under age five died of cancer each year. By 1985 that number had been reduced to 700, suggesting that two-thirds may be permanently cured (figure 18). Several other cancers of the young have also become curable; the number of cancer deaths of people under thirty has fallen from 10,000 to 7,000 a year. In contrast, mortality from the most frequent cancers afflicting older people has hardly dropped in the last twenty-five years.[48] This fact represents the severest challenge to research and the pharmaceutical industry.

One of that industry's outstanding recent successes in another field has been the development of beta blockers, the first of which was discovered by James W. Black in Britain. Some beta blockers relieve angina and reduce blood pressure; another (also developed by Black) makes gastric ulcers regress by inhibiting the secretion of hydrochloric acid. Pharmacologists have always tested new drugs on animals before trying them on people, and this procedure was believed to safeguard patients from possible toxic effects. The thalidomide tragedy, which happened in 1962, shook the public's confidence in the pharmaceutical industry and led to a tightening of safety regulations in most countries.

The administration of these regulations has become cumbersome in many countries, especially the United States, which used

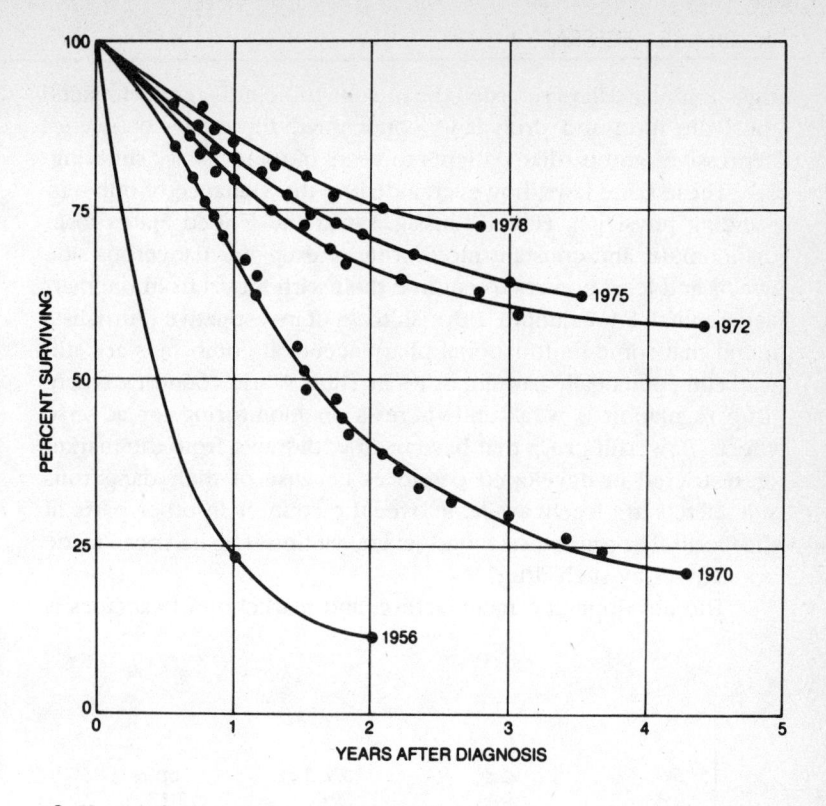

18. Years of survival for children with leukemia, 1956–1978. Chemotherapy has made leukemic children survive longer. In 1956 only 10 percent were still alive two years after the diagnosis; by 1978 that number had risen to 75 percent. SOURCE: Denman Hammond at the Children's Cancer Study Group, in John Cairns, "The Treatment of Diseases and the War against Cancer," *Scientific American* 253 (November 1985): 31–39.

to be the country from which most new drugs originated. There the time taken from the patenting of a new compound to its marketing averaged three years in the early 1960s, rose to seven and a half years in the early 1970s, and was at nine years in 1978–79, largely as a result of the ever more elaborate trials and safety tests that are demanded. (Figure 19 shows parallel time spans in Great Britain.) For instance, the efficacy of lithium carbonate against manic depression was discovered in the 1950s; by 1960 the drug was in general use in Europe, but no American firm considered it economical to undertake the elaborate tests required by the Food and Drug Administration because lithium carbonate is a simple inorganic compound that cannot be patented and sold with exclusive rights. Similar delays retarded the introduction of the beta blockers; thus, the food and drug laws condemned thousands of manic-depressives and cardiac patients to years of unnecessary suffering.

These same laws, however, aided by the vigilance of one outstanding physician, Helen Taussig, saved the United States from thalidomide; and constant monitoring of drugs for dangerous side effects and legal powers to enforce their withdrawal from the market are vital. For example, a British team of investigative journalists found that some multinational pharmaceutical companies are still marketing potentially harmful drugs in Third World countries, where drug regulation is weak and there is no monitoring for adverse effects. Powerful drugs that have been withdrawn from the market or restricted in developed countries because of their dangerous side effects are freely available over the counter in other parts of the world. The journalists found several victims who had been made critically ill by such drugs.[49]

The development, manufacture, and marketing of vaccines is

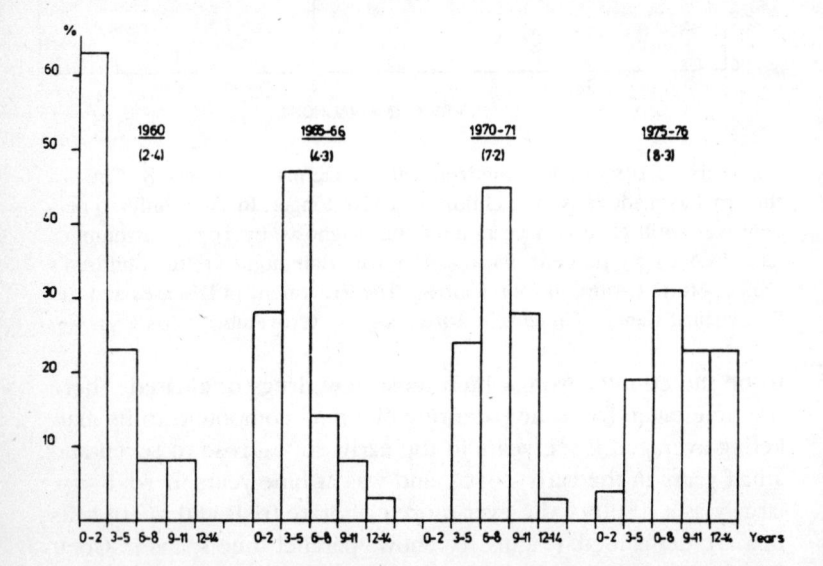

19. Interval between the first patent and release of new drugs in Great Britain, 1960 to 1975–1976. The figures in brackets give the average interval. SOURCE: M. F. Steward, "Public Policy and Innovation in the Drug Industry," in *Proceedings of Section 10 (General) of the British Association for the Advancement of Science, 139th Annual Meeting, 1977,* eds. Douglas Black and G. P. Thomas (London: Croom Helm, 1980).

said to be discouraged by the enormous damages awarded when vaccines accidentally injure people, even though such accidents are very rare. Vaccines do not bring in the large profits that patented drugs do, and the cost of litigation, if added to that of the product, can force vaccines off the market. In consequence, the number of vaccine manufacturers has been dropping steadily, a decline that presents a greater danger to public health than the rare accidents. It would be better if there were a system of no-fault compensation for the victims when no negligence can be proved.

Tighter legal restrictions have come at a time when it is becoming intrinsically harder to discover new drugs; an average of 7,000 organic compounds now have to be synthesized before a pharmacologically useful one is found. Compare this with the mere 605 arsenicals Paul Ehrlich had to make before he found his magic bullet against syphilis. As a result, the cost of putting a new drug on the market rose fivefold in real money terms between 1960 and 1975 and is now on the order of $50 million. The cost of chemistry and pharmacology make up no more than one-third of this huge total; the remainder is spent on toxicity and clinical trials and on other developments needed before a license can be applied for. The number of chemically new drugs put on the market is falling, and the proportion of money being spent on development continues to rise at the expense of research (figures 20–22).[50]

In the United States pharmaceutical literature is spiced with vitriolic attacks on the Food and Drug Administration for killing progress, while the administration retorts that its controls have mainly reduced the issues of ineffective new drugs, and that increased costs and development time have other causes.

New drug research is expensive, but great advances in the public health of developing and developed countries could be made easily and cheaply by teaching people to apply existing knowledge. Vulimiri Ramalingaswami of the Indian Medical Research Council has pointed out that in his country many diseases—such as goiter, xerophthalmia, and nutritional anemia—are still endemic, even though they could be prevented very cheaply with iodized salt, vitamin A, and iron sulphate. Death from infantile diarrhea is frequent, although it can easily be prevented by oral rehydration with a glucose-salt mixture. Supplementary feeding programs for infants make little impact unless mothers are also taught how to prevent malnutrition and illness in young children.

To bring such medicine and simple knowledge to the people,

the government of India has launched a rural health scheme, which should provide every village with a population of over 1,000 people with a community health worker trained for three months in a neighborhood primary health center. By 1985 there were to be one male and one female multipurpose health worker for every 5,000 people in rural areas. The scheme is similar to that of the successful barefoot doctors in China and counteracts the tendency of doctors trained in hospitals to be more interested in setting up prestigious intensive care units in towns than in organizing elementary medical services for the rural population.[51]

The World Health Organization has come to realize that large strides in public health could be made in many countries by adopting the elementary hygiene that was developed in Europe before the advent of modern medicine (figures 12 and 14). It has therefore launched a campaign to provide every human being with clean water and sanitation by 1990.

The United Nations Children's Fund, UNICEF, reports that in poor countries preventable infectious diseases are still killing mil-

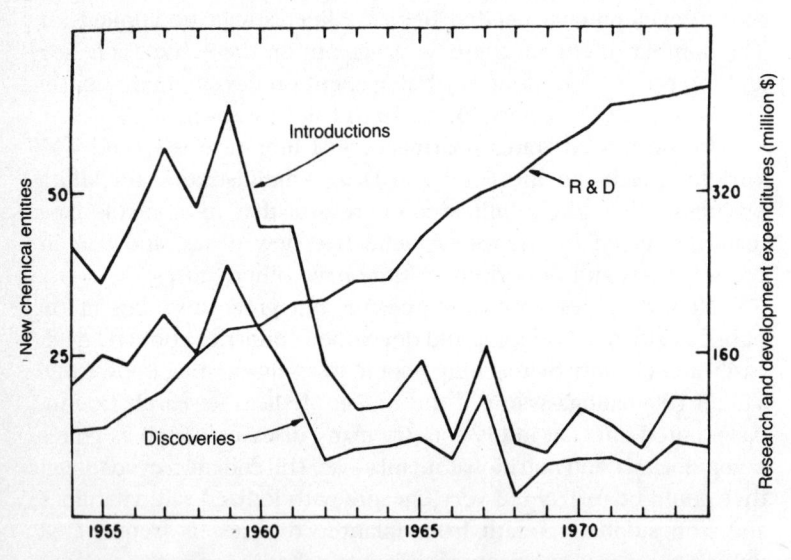

20. Discovery and introduction of new drugs compared with the costs of drug development in the United States in 1958 dollars. SOURCE: H. G. Grabowski, J. M. Vernon, and L. G. Thomas, "Estimating the Effect of Regulation on Innovation: An International Comparative Analysis of the Drug Industry," *Journal of Law and Economics* 21 (1978): 133.

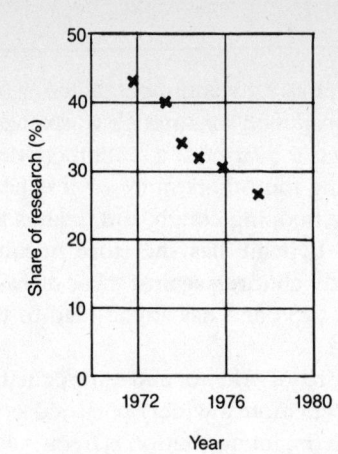

21. Percentage share of research in the research and development budget of Hoechst A.G., 1972–1980. SOURCE: *Arzneimittelforschung in Deutschland,* Pharma, Bundesverband der Pharmazeutischen Industrie, Karlstrasse 21, 6000 Frankfurt, 1979–1980.

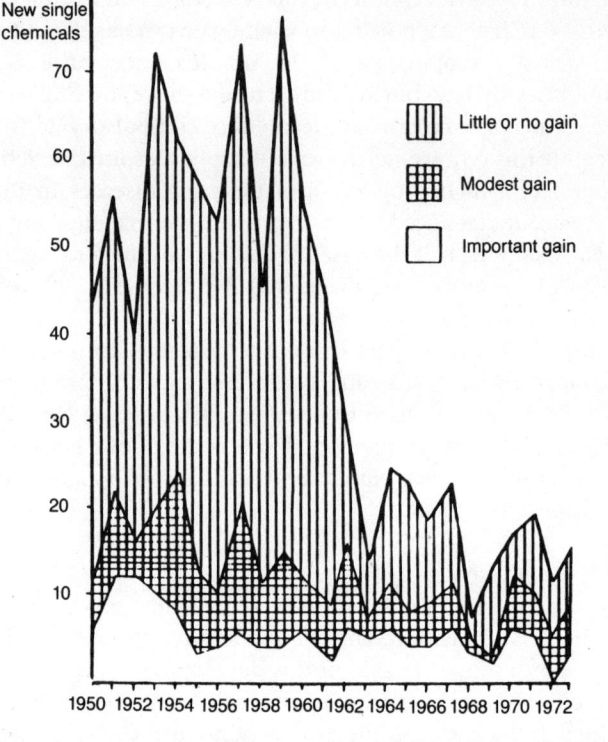

22. Annual approval of new drugs in the United States and their efficacy, 1950–1972. SOURCE: M. F. Steward, "Public Policy and Innovation in the Drug Industry," in *Proceedings of Section 10 (General) of the British Association for the Advancement of Science, 139th Annual Meeting, 1977,* eds. Douglas Black and G. P. Thomas (London: Croom Helm, 1980).

lions of children or leaving them physically and mentally crippled. Vaccination against polio was introduced in America thirty years ago, but worldwide polio still affects a quarter of a million children annually, killing 25,000 and leaving the others more or less paralyzed. Polio, measles, diphtheria, whooping cough, and tetanus kill about 4 million children a year. UNICEF has therefore begun a campaign to vaccinate all the world's children against these diseases by 1990. The campaign began in 1985 and has already led to the vaccination of 18 million children.

You need not go to the Third World to find unvaccinated children, because compulsory vaccination is widely regarded as an infringement of civil liberty. In Britain, immunization is freely available under the National Health Service, but one-fifth of children are not immunized against diphtheria, tetanus, polio, and German measles, nearly one-half are not immunized against measles, and over half not against whooping cough. In America, vaccination is also not compulsory by law, but schools refuse to accept children who are not vaccinated, which makes it in effect compulsory; in Britain unfortunately this requirement does not apply. Despite the National Health Service, a high proportion of treatable diseases in British adults remain undiagnosed, or if diagnosed are inadequately controlled. For example, half the cases of diabetes are never diagnosed, and in half of those that are diagnosed the disease is not kept in check.[52]

In summary, a large part of untimely death and disability in both developed and developing countries could be prevented if there were a public will to act on existing knowledge. New drugs and treatments cannot be developed unless the public is willing to accept a measure of risk, since zero risk can be bought only at infinite expense—if at all.

LONGEVITY IS EXPENSIVE

Figure 13 showed the increase in average life expectancy that Western hygiene, nutrition, and medicine have brought to Japan. Figure 23 shows the other side of this development: the enormous rise in medical costs incurred in the care of increasingly aging populations. In Switzerland medical expenditure per head of population is rising one and a half times faster than prices and two and a half times faster than wages. Between 1949 and 1983, the average annual cost of the prescriptions handed out by a general practitioner in Britain quadrupled at constant prices. In the United States medical

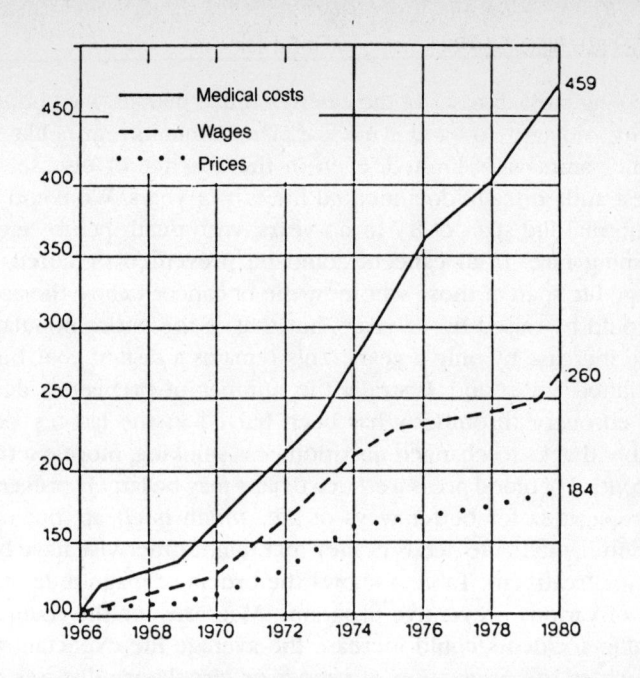

23. Consumer prices, wages, and medical costs per insured person in Switzerland, 1966–1980. Medical costs have risen more steeply than prices and wages, partly because of the aging of the population. The numbers on the left give prices relative to their level in 1966 which is taken as 100. SOURCE: Professor M. Schär, Zurich.

costs trebled in ten years and in 1983 consumed 11 percent of the gross national product. These rises are causing governments concern, especially since the demand for medical treatment always exceeds what is available from private insurance or the state.

Science is criticized for forcing ever more sophisticated and expensive diagnostic tools on the medical establishment, but misdiagnoses leading to wrong treatment may cause greater expense. For example, in Britain one asks whether it pays to replace old people's arthritic hips at public expense. The alternative would be that patients become bedridden and have to be nursed, which costs more. By contrast, it would be much cheaper if scientific research led to the discovery of the cause of arthritis and to its prevention or cure without surgery. Such success was achieved by research on gastric ulcers, which need no longer be removed surgically because they can now be kept in check by James Black's drug cimetidine.

The ideal would be reached if scientific research succeeded in preventing most diseases of the elderly, so that people "died healthy" at a ripe old age. How old is unclear. The human life span, like that of other animals, is limited, even in the absence of disease. The longest authentically documented life is 114 years. We could aim at a normal life span of 85 to 90 years, with death before age 70 becoming rare. If all cancers could be prevented or cured, the average life span of those who now die of cancer before the age of 65 would be raised by 12 years, but that of the entire population would increase by only 2 years. This remains a distant goal, but in the United States and Australia the number of premature deaths from coronary thrombosis has been halved in the last 15 years, possibly thanks to changed nutrition, less smoking, more exercise, and control of blood pressure. Such deaths may be largely prevented by propaganda for better ways of life, which is cheap, but often preventive medicine needs money that might otherwise have been used for treatment. Table 4 shows the orders of magnitude of the costs of various preventive programs. Measures for the reduction of traffic accidents could increase the average life expectancy by as much as the prevention of cancer or circulatory diseases, and these measures are among the cheapest. The cost of years of life saved by early diagnosis goes up in inverse proportion to the fraction of diagnosed cases, hence the rise from $1,175 for the first test for intestinal cancer to $47 million for the sixth, when new cases would be discovered very rarely. Prevention may always be better than cure, but it is not necessarily cheaper. For example, it turns out to be cheaper to operate on people with narrowed aortas than to screen the entire population for the first signs of this condition. In view of these costs, it is unlikely that even the richest countries will be able to afford to screen most of their population for the earliest signs of the most frequent cancers and other common diseases, nor is screening always certain to prevent these diseases. Besides, frequent screening may be injurious and give rise to neuroses.

Would prevention of early deaths increase the number of physically and mentally handicapped old people to a financially unbearable extent? Doll and Peto argue that this does not follow.

It is obvious from international comparisons of age-specific, disease-specific death rates that the majority of deaths in middle age could be prevented. It is similarly obvious that the major diseases that cause disability in old age now, such as Alz-

TABLE 4

COST OF PREVENTIVE TREATMENT OR RESEARCH
FOR VARIOUS CAUSES OF DEATH PER LIFE-YEAR SAVED

	PREVENTIVE MEASURES OR RESEARCH	COST PER LIFE-YEAR SAVED ($)
Cancer, general	Study of possible protection by beta-carotene	100
Intestinal cancer	First test for blood in stools	1,175
	Sixth test for blood in stools	47,000,000
Breast cancer	Annual X-ray mammography	c. 5,000
Kidney failure	Hemodialysis	30,000
Coronary occlusion	Surgical shunt	4,000
	Screening of all middle-aged men for exercise tolerance and thallium tests	22,000–35,000
	Cholesterol screening of children	3,400–6,400
Heart failure and stroke	Screening and therapy for high blood pressure	33,000–64,000
Automobile accident	Improved construction of cars, safer traffic regulations, and so on	2,000

SOURCE: Richard Doll, Richard Peto, David Evered, Julie Whelan, eds., *The Value of Preventive Medicine,* CIBA Symposium No. 110 (London: Pitman, 1985).

heimer's disease, changes in cerebral blood flow, and rheumatoid arthritis, will eventually be as preventable (or treatable) as the major cancers or vascular diseases. This is because there is no systematic reason why the reduction of the age-specific death rate for one disease should necessarily cause an increase of that for another. The prevention of one type of cancer *might* inadvertently cause an increase in another, but it might just as well cause a decrease; there is in general no systematically positive correlation between one disease and another, among people of a given age. ... If medicine can prevent diseases that have high fatality rates, why shouldn't it prevent diseases with low fatality rates? We may find that some of these diseases are the automatic results of ageing which we cannot reverse, but why be pessimistic about it? There is nothing in past history that indicates that we have to be.[53]

GENE TECHNOLOGY AND MEDICINE

Gene technology has existed for only fifteen years but has already found some useful medical applications. I explain in a previous section on "Biotechnology and Genetic Engineering in Agriculture" why no one thinks of fiddling with the human genome; however, that genome can now be studied in minute detail. In principle, any human gene can be isolated, cloned, and deciphered, and the protein coded for by that gene can be made in any quantity needed for medical purposes.

Inherited diseases are responsible for about a third of all children's admissions to hospitals and for nearly half of all deaths of children under age fifteen. Gynecologists and molecular biologists have worked out methods of prenatal diagnosis that have brought about drastic reductions in the number of children born with one of the severest and most common inborn diseases.[54] This is thalassemia major, an anemia that is frequent around the Mediterranean and in Southeast Asia. Its heterozygotic carriers are healthy, but one in four of the children of two carriers is likely to be born with the anemia. Simple chemical analysis of the hemoglobin in the red blood cells or of the DNA contained in white blood corpuscles can tell whether the parents are carriers. If they are, a tiny fiber snipped off the membrane surrounding the eight- or nine-week-old embryo provides enough DNA to diagnose if that embryo has inherited thalassemia at all, if it has inherited it from only one parent and will be born healthy, or if it has inherited it from both parents and will be afflicted by the disease. If the embryo has inherited the disease from both parents, they can decide if they want to terminate the pregnancy and try to conceive a healthy child.

Bernadette Modell is a pediatrician who experienced the distress caused to Cypriot families in London by the raising of severely crippled thalassemic children. With the help of expert colleagues

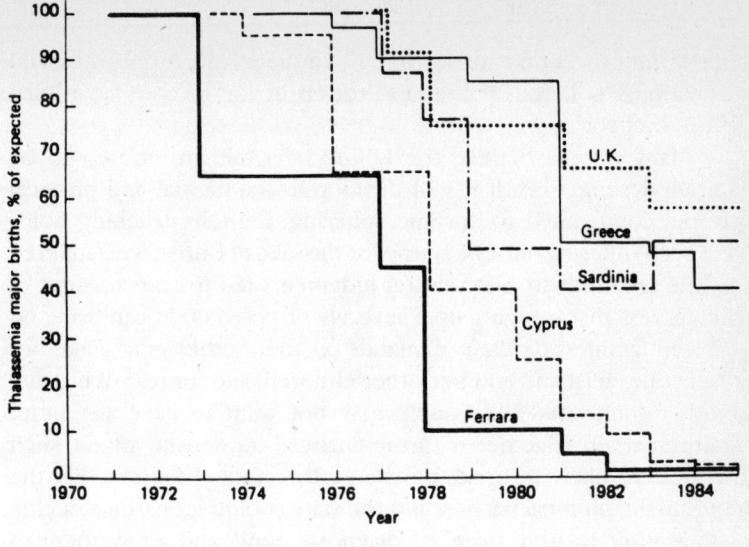

24. Reduction in the number of births of thalassemic babies as a result of antenatal diagnosis in the Italian city of Ferrara, on the Mediterranean islands of Cyprus and Sardinia, in Greece, and in the United Kingdom. The numbers are percentages relative to those expected on the basis of the frequency of thalassemic births in 1970, taken as 100. SOURCE: *Report of the World Health Organization European/Mediterranean Working Group on Haemoglobinopathies,* Brussels: 14 March 1986; Paris: 20–21 March 1987.

she organized prenatal diagnosis at University College Hospital. Her clinic was soon beleaguered by pregnant Cypriot women, and her work proved so successful that doctors from Mediterranean countries came to be trained, so that they could introduce prenatal diagnosis in their own countries. By 1983 its practice had reduced the number of thalassemic children born annually in Cyprus from 70 to 2, in Greece from 300 to 150, in Sardinia from 70 to 30, and in the Italian city of Ferrara from 25 to 0. In Italy as a whole the number has dropped by 60 percent and in Sardinia by 70 percent (figure 24).[55]

Note that in 1984 the incidence of thalassemic babies born in Sardinia rose again. This happened because health authorities failed to finance the necessary education campaign in remote villages. In the United Kingdom and other Northwest European countries there has as yet been little reduction in the number of thalassemic babies born because the disease is endemic only among recently arrived ethnic minorities that are scattered over a few industrial cities, and no national policies for the eradication of the disease have been put into effect.

Many people believe the killing of a human embryo to be morally wrong, even if it will develop into a mental and physical cripple condemned to chronic suffering. Catholic teaching holds that such suffering must be borne for the sake of Christ, who suffered for the sake of man. Such moral judgments fail to take account of the distress that bringing up a severely crippled child can bring on afflicted families. Its great demands on the mother may make her neglect her husband and her other children, and for fear of bearing another such child the couple may not want to have any more children at all. Together with the financial burdens involved, such stresses are likely to break up the entire family. I believe that the church, the political parties, and the state should let parents decide if they want to have prenatal diagnosis done, and allow them to take the responsibility for termination of pregnancy from informed choice.

Perhaps the strongest argument in favor of such a policy comes from the observation that couples who have had one thalassemic child and then do not resort to antenatal diagnosis tend to have no more children, while couples who *do* make use of it tend to have more and healthy children. For example, Cypriot couples who have had one thalassemic child and do not resort to antenatal diagnosis have had on average one child per forty-seven years of marriage, while those who do, have had on the average of one healthy child per 4.6 years of marriage.[56]

In the United States, 70 to 80 babies with thalassemia major and about 1,000 black babies with sickle-cell anemia, another severe blood disease, are born each year. The numbers born with thalassemia major are going down, partly thanks to prenatal diagnosis and partly because of increasing intermarriage of Greeks and Italians with other ethnic groups.[57] There is no evidence yet of a decrease in the number of sickle-cell anemia babies born, perhaps because information about prenatal diagnosis has not yet spread in the black community. The Mediterranean experience shows that an educational campaign linked to prenatal services could drastically reduce the number of babies born with sickle-cell anemia and raise the number of healthy babies born to sickle-cell carriers.

Not all inherited diseases can as yet be diagnosed in unborn babies. Down's syndrome, spina bifida, and cystic fibrosis can. Hemophilia and certain muscular dystrophies can also be diagnosed, but they often arise by new mutations in children of normal parents and become evident only after birth.

Will gene therapy cure inherited diseases? Most inherited diseases are the result of mutations that disturb or abolish the function of an essential protein. To help the patients, either the missing protein must be introduced (for example, the missing clotting factors of hemophiliacs are introduced into the blood by injection) or the gene needed to code for the missing protein must be introduced in such a way that it is synthesized in the patient's tissues. For this method to succeed, the gene must first be spliced into the patient's chromosomes, since otherwise it would soon be lost or degraded. This is now being attempted in the following way. The missing gene is spliced into the chromosome of a virus. When this virus infects the patient, it transfers its entire chromosome into the nuclei of some of the patient's cells. These cells should then synthesize the missing protein. To avoid their becoming infected with the virus, the genes needed for the virus's replication and pathogenic effects are excised from the viral chromosome. The entire procedure is difficult and uncertain, because transfer of the viral DNA into the nucleus of a human cell is a haphazard process, and there is as yet no way of ensuring that the missing gene is spliced exclusively into the correct position on the correct human chromosome rather than anywhere else or on any of the forty-five others; though this has already been done in a mouse.

Several American medical scientists are trying this method on children suffering from two of the worst and most intractable inherited diseases: adenosine deaminase deficiency, which paralyzes the immune system, and Lesch-Nyhan syndrome, which makes children mutilate themselves, for example, by chewing off their fingers. The scientists extract a little of these children's bone marrow, incubate it with the virus carrying the missing genes, and then return these "transformed" cells to the children's marrow. The viral infection is thus introduced not in the patient but in vitro in cells taken from the patient, and only after the virus has been rendered harmless. Once back in the bone marrow, these transformed cells are supposed to multiply and supply the patient with the missing protein.

This procedure resembles that used in bone marrow transplantation for leukemia, in which the patient's cancerous marrow is first killed by a "lethal" dose of radiation and then replaced by healthy marrow from a donor. It is therefore unjustified to condemn these attempts as tinkering with the human genome, as opponents of genetic engineering have done. They do run the remote risks that the defective viral chromosome might become infectious by combining with other, undetected viral chromosomes or with parts of the patient's own DNA, or that it might accidentally activate one of the patient's cancer genes. For these reasons the treatment is being tried only on very sick children. So far the attempts have been unsuccessful, because the newly introduced genes have failed to stimulate the synthesis of significant quantities of the missing protein.

An alternative approach, which has succeeded in some cases, is the transplantation of bone marrow from a healthy sibling. Scientists have recently made a mouse model of Lesch-Nyhan syndrome, consisting of embryonic cells that lack the gene for the enzyme which is deficient in that disease. They treated these cells with DNA that included the gene and was otherwise constructed in such a way that it would home in exclusively on its correct target, the position on the mouse chromosome where that gene was absent. This targeted transfer was successful in about one in a million of the treated cells and raises hopes that a similar transfer might eventually work in human bone marrow cells.[58]

Some inherited diseases, such as Huntington's chorea, are caused not by the absence or nonfunction of a protein but by the pathogenic actions of an abnormal protein, the product of a sick gene. In theory, such a gene can be inactivated, but so far no practical way of doing it has been found.

Gene therapy, if it succeeds, is bound to be expensive to begin with, but it may soon become cheaper than lifelong care of patients who suffer from these disorders. The choice of the children to be treated, the consent of their parents, and the financing of treatment are likely to raise difficult ethical problems. In America these have been considered by a working party consisting of three natural scientists, three clinicians, three specialists in ethics, three lawyers, two political scientists, and one layman. Their shrewd and humane conclusions have been published and could serve as a model for other countries facing these problems.[59] Gene therapy of the fertilized human egg is out of the question, since it succeeds in only a fraction of the treated eggs, while many might end as abnormal births.

Gene technology has brought the greatest progress to cancer

research since 1910, when Peyton Rous discovered the first cancer virus in chicken. Molecular biologists have discovered that responsibility for the chicken cancer rests with a single gene, which is transferred from the chromosome of the virus to one of the host chromosomes after infection. The same applies to other cancer-producing viruses of animals and birds.

At first these studies were thought to be rather academic, because few human cancers were known to be of viral origin, but this line of research soon led to a discovery that has brought us close to an understanding of the molecular basis of some of the major human cancers. Molecular biologists have found genes in normal human chromosomes that closely resemble viral cancer genes. Spontaneous mutations can turn these normal genes into cancer genes, and the positions where these mutations occur coincide with those that differentiate the normal human genes from the viral cancer genes. The functions of these normal human genes are not yet clear, but many of these genes appear to control the synthesis of proteins that either stimulate cell division or are the receptors for proteins that do so. Mutations of these genes may therefore allow cell division to go out of control. There is now hope that we shall soon know the exact molecular mechanism that turns normal cells into cancerous ones. This knowledge may not lead directly to better treatments for cancer, but it is the first requisite for better treatments. Until now cancer research has been groping in the dark.[60]

Gene technology may soon result in better treatment of coronary thrombosis. Human tissue contains minute quantities of a protein called plasminogen activator (TPA), which initiates the dissolution of blood clots. Biochemists have isolated the gene that codes for TPA's synthesis, cloned it, and introduced it into coli bacteria, yeast, or mammalian cells that can be grown in culture and made to produce the protein in quantity. TPA is being made by several gene firms and is likely to be widely used for the dissolution of blood clots in coronary and other thrombosis as soon as it has been cleared by the Food and Drug Administration.[61] These same firms are also trying to make, in coli bacteria or yeast, the clotting factor that hemophiliacs lack; now this substance is isolated from donated human blood, which may carry viral infections. Unfortunately these efforts are beginning only after the tragic infection of many hemophiliacs with the AIDS virus; in Great Britain three-quarters of the hemophiliacs are said to have been infected by preparations of clotting factors imported from the United States.

Karl Landsteiner's discovery of the blood groups early in this century made it possible to disprove parentage of children but not to prove it, and laws have been formulated accordingly. The English geneticist Alec Jeffreys recently discovered that human chromosomes contain certain segments of DNA whose nucleotide base sequences are distinct in each individual. Chemical analysis of these segments provides a characteristic pattern of lines, inherited equally from father and mother. If half the lines of a child coincide with half the lines of the disputing father, then the possibility of anyone else being the father can be safely excluded.[62] In forensic cases, analysis of the DNA extracted from traces of blood, skin, or semen allow the guilt of the accused to be proved or disproved with certainty. In England this recently led to the proof that a man convicted and imprisoned for murder and rape was innocent.

Genetic engineers' most urgent task is the search for a vaccine to stem the AIDS epidemic. Vaccination against the polio virus became possible only after the American biologist J. F. Enders had discovered how to grow the virus in cultures of embryonic chicken cells. The AIDS virus was first isolated by Luc Montagnier at the Pasteur Institute in Paris in 1983; it grows only in specialized human blood cells (T lymphocytes), which cannot be cultured on a large scale. Besides, this culturing would be too dangerous, and gene technology opens an easier way.

Virologists and molecular biologists found the virus to be enveloped by sixty identical copies of a coat protein. On infection with the virus, this is the protein that the immune system would "see" first and that should be capable of stimulating the system to produce antibodies against the virus. Scientists have isolated the gene for this protein, cloned it, and spliced it into the chromosome of the baculovirus. Larvae infected with that virus produced the coat protein in large quantities, but animal experiments showed it to be only weakly antigenic. W. F. H. Jarrett at the University of Glasgow raised the virus's antigenicity a thousandfold by adsorbing it onto a carbohydrate particle extracted from the bark of a South American tree that is shaped like a small spherical virus, a new technique introduced by Swedish immunologist B. Morein and others. On its surface the immune system "sees" many viral coat protein molecules side by side, as in a live virus, and therefore reacts more strongly. Another vaccine, developed in the United States, consists of vaccinia virus that carries, in addition to its own genes, the gene for the coat protein of the AIDS virus and should therefore have its surface lined with many copies of that protein. Such vaccines have been made against the Simian (monkey) Immunodeficiency Virus which resembles the human one. They elicited antibodies, but these antibodies failed to protect the monkeys from subsequent infection with the virus. The reasons for this failure are not understood and so far there is no AIDS vaccine in sight.

Gene technology is leading to the development of safe vaccines against malaria and other tropical diseases, to the benefit of public health in many parts of the world.

HAS SCIENCE DEHUMANIZED MEDICINE?

Despite its spectacular successes many people have become disillusioned with modern medicine and accuse science of having dehumanized it. The roots of this accusation go back to past medical practice, as it is described by Lewis Thomas in his autobiography *The Youngest Science*. He grew up as the son of a general practitioner and surgeon in a small New England town and became a student at the Harvard Medical School in the 1930s.[63]

When Thomas's father took him on his rounds, he told his son how little he was able to do for the many people who asked for his help. Most diseases killed some people and spared others, and if you were one of the lucky ones you were convinced that the doctor had saved you. The medicines Thomas's father carried in his bag were placebos, or tonics, or useless mixtures such as elixir of iron, strychnine, and quinine. When young Lewis Thomas entered medical school, he was taught to diagnose diseases by their symptoms and manifestations in the laboratory; treatment formed only a minor part of the curriculum. He was taught that what ill patients wanted to know was the name of their disease, its possible cause, and its likely prognosis. When he became an intern, he and his colleagues began to realize that they could do nothing to change the course of most of the diseases they diagnosed. "Whether the patient survived or not depended on the natural history of the disease itself. Medicine made little difference."

William Osler had been the great teacher of classical diagnosis. He was Professor of Medicine at The Johns Hopkins University in 1900 and later became Sir William Osler, Regius Professor of Medicine at Oxford University. René Dubos writes of him that "To the end of his life, he remained unshaken in his belief that medicine can be learned only at the bedside, and that its most important aspect is the art of establishing the right kind of rapport between physician and patient." According to Osler, "Faith in the gods or the saints cures one, faith in little pills another, hypnotic suggestion a third, faith in the common doctor a fourth."[64] When I arrived in England in 1936, a doctor's bedside manner was regarded as his supreme asset.

The great advances of the past fifty years have shifted the emphasis of medical teaching toward scientific methods of diagnosis and treatment, sometimes to the neglect of the personal relationship between doctor and patient that existed in the past. As a result, doctors may diagnose the disease, but fail to discover the cause that only personal knowledge of the patient, such as the old-style family doctor possessed, could reveal. Besides, the plethora of gadgets that are brought to bear on patients can make them feel as if they are machines being probed by engineers. Medical schools have reacted to these misgivings by making changes in their teaching; general practitioners are enlisting the help of trained counselors to relieve the traumas that underlie some of their patients' symptoms; and hospitals employ social workers or physicians trained in social work to discover some of their distressed patients' personal histories. These measures will go some way toward restoring the balance between the old medicine and the new, but conditions that fail to respond to scientific treatment will always remain and can sometimes be relieved by faith in a healer.

SCIENCE AND ENERGY

SOURCES OF ENERGY

Our civilization is based on cheap, readily convertible energy, mainly from coal, oil, and natural gas, which are being used at an increasing rate (figure 25). Coal and oil are also the raw materials for many essentials of modern life made by the chemical industry. How long will they last? Figure 26 shows that their abundance marks no more than a moment in man's history.

Today oil supplies nearly half of the world's energy. Figure 27 shows a recent estimate of future world oil production, by Sir Peter Baxendell, managing director of the Royal Dutch/Shell companies. According to him, the present production rate of about 50 million barrels a day might be maintained with oil from known reserves in existing fields and from fields still to be discovered only until about

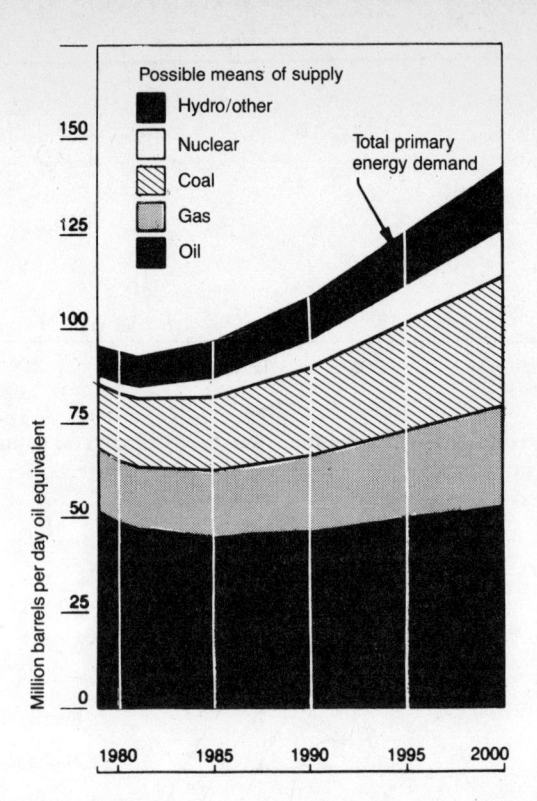

25. World energy demand and possible means of supply, 1980–2000. (Excludes USSR, Eastern Europe, and China.) SOURCE: U.S. Department of Energy, Energy Information Administration, *Annual Energy Outlook, 1984*. DOE/EIA = 0383 (84).

2015. Further supplies depend on enhanced recovery from existing fields, from which no more than a third of the oil is currently being extracted because recovery of the remainder requires techniques such as high pressures of nitrogen, carbon dioxide, or steam, which are uneconomic at present prices and which will require many years of research and pilot testing before they can be applied routinely in the field. The same is true of the extraction of oil from tar sands, of which there are large reserves in Alberta, Canada.[65] If oil from all these reserves could be extracted at reasonable prices, the reserves might last for the lifetime of our grandchildren. They may last longer only by becoming prohibitively expensive.

Natural gas now supplies about one-fifth of the world's energy demand. Those resources that are now being tapped, such as North Sea gas, will not last long; large untapped reserves exist in the Middle

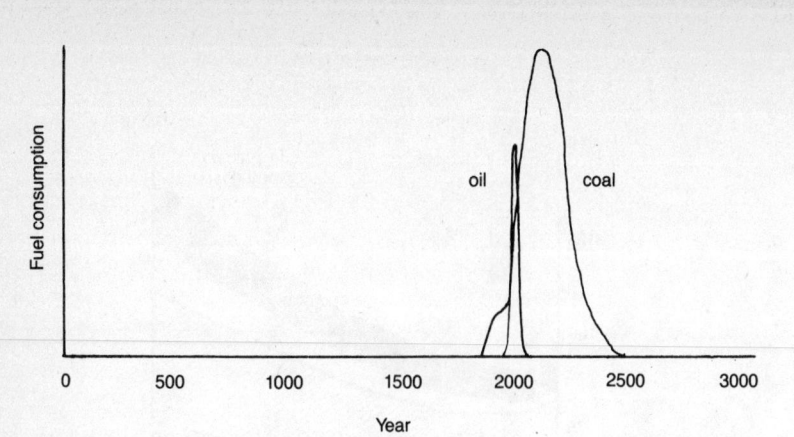

26. Expected duration of fossil fuels, A.D. 0–3000. Oil and natural gas will last but a moment in man's history. SOURCE: Sir George Porter, president of the Royal Society, London.

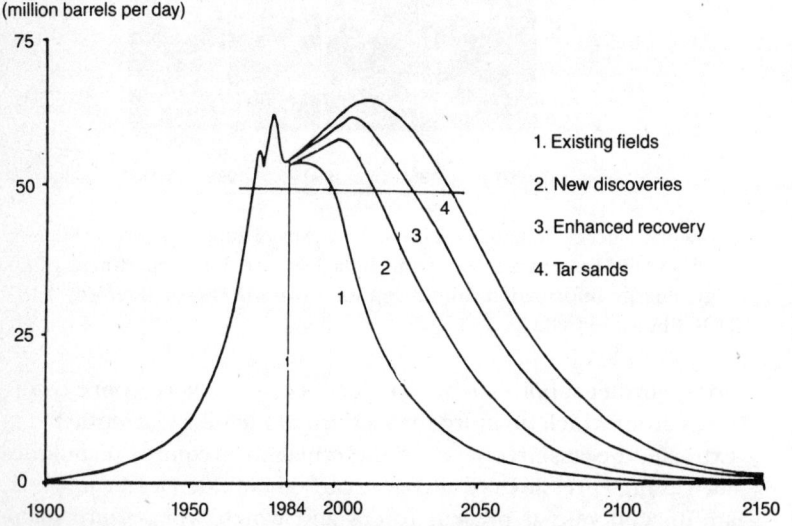

27. Forecast of world oil production by Sir Peter Baxendell, managing director of Royal Dutch-Shell. The curves indicate ratio of production in millions of barrels per day: (1) from existing fields; (2) from new fields likely to be discovered; (3) by recovery from existing fields and new fields of oil that require expensive new techniques which are uneconomic at present prices; and (4) recovery of oil that is absorbed in tar sands and is uneconomic to extract at present prices. Alberta, Canada has large deposits. The horizontal line shows 1984 daily consumption. SOURCE: Peter Baxendell, "Enhancing Oil Recovery—Making the Most of What We've Got," *Transactions of Mining and Metallurgy* 94A (April 1985): A84–A89.

East, Africa, and the USSR, but gas from them will be expensive to transport to North America and Western Europe. Coal supplies about a quarter of the world's energy, and more than half of the coal mined is used to generate electricity. Almost 90 percent of known coal deposits lie in only four countries: the USSR (42 percent), the United States (26 percent), China (13 percent), and Australia (6 percent). These deposits contain about 250 times the coal produced in 1980.[66]

The world's total reserves of fossil fuels are estimated to amount to the equivalent of 5.5 trillion tons of coal. If the world population were to stabilize at 10 billion and the average consumption per head remained the same, the reserves would last 275 years, but this projection assumes that the standard of living in the developing world would not rise significantly. If it did, more energy would be consumed, and the fossil fuels would run out sooner.

NUCLEAR REACTORS

If we want to preserve civilization for our descendants we have to find other energy sources. Nuclear fuels could cover the growing energy demand much longer and also help us conserve our precious fossil fuels. There are two kinds of nuclear reactor: thermal and fast breeder.[67] Thermal reactors use either natural uranium, which contains 99.3 percent nonfissile uranium 238 and 0.7 percent fissile uranium 235, or natural uranium enriched with uranium 235. They have to be refueled because the number of fissile nuclei of uranium 235 used in the process is greater than the number created by neutron capture. Since the world's uranium resources are limited, the supply of uranium 235 may become scarce in the lifetime of our children.

A thermal reactor breeds fissile plutonium from nonfissile uranium 238, but the amount of plutonium it breeds is less than the amount of fissile uranium 235 it burns. By contrast, a fast breeder reactor breeds more plutonium in its surrounding blanket of uranium 238 than it burns in its core, and this plutonium can be extracted and used to refuel the core or to build more reactors. Because fast breeder reactors employ nonfissile uranium 238 as a fuel, their use of natural uranium is sixty times more efficient than that of thermal reactors and would make the world's uranium supplies last correspondingly longer.[68] For a generating capacity of 1,000 megawatts of electricity, enough for an industrial town of about a million inhabitants, such a reactor would have to contain a core made up of 5 metric tons of plutonium mixed with 20 metric tons of uranium, surrounded by a blanket of uranium, which could be the spent fuel of thermal reactors from which the plutonium has been extracted. Such plutonium is by now available in excess.

In England, a fast breeder reactor with a maximum load of 250 megawatts had teething troubles, but by 1984 these had been overcome, and since then it has worked well with a 60 percent load. France began the construction of a breeder reactor of 1,200 megawatts, the Superphénix, near Lyons in 1976. It was connected to the French electricity network in 1986 and now runs at full capacity. West Germany also built such a reactor in collaboration with Dutch and Belgian companies, but it has not been licensed yet. Incidentally, such reactors are called *fast* because they use fast neutrons, not because they breed plutonium quickly.

A hazard of fast breeders is the plutonium trade they entail. Since their plutonium has to be extracted initially from the spent fuel of thermal reactors, plutonium would become an international commodity. A few kilograms of plutonium are enough for an atomic bomb, yet 5,000 kilograms or more are needed for a reactor. This difference opens up frightening possibilities that plutonium could fall into the hands of terrorists.

Another hazard, common to both fast breeder and thermal reactors, is the radioactive waste generated in their cores and casings. It is not easy to appreciate the colossal magnitude of this problem. Figure 28 shows the amount of radioactivity produced per unit of electricity generated and its decay over the years, but it is important also to realize the volumes involved. At each of the larger thermal reactors, about 100 cubic meters of solid waste accumulate each year. In 1975, 12,000 cubic meters of solid waste, containing half a metric ton of plutonium, were stored at the British reactor in Sellafield. In addition, 600 cubic meters of highly radioactive liquid waste were stored there, and ten times that amount is expected by the year 2000.

In Britain, no definite plans for the eventual disposal of this waste have yet been decided on, although some technical solutions have been formulated. For example, freezing the waste in glass and burying it in stable rock formations, either on land or under the seabed, is under discussion. The Royal Commission for Environmental Pollution expressed anxiety about the absence of a fully worked out solution and recommended that "there should be no commitment to a large programme of nuclear fission power until

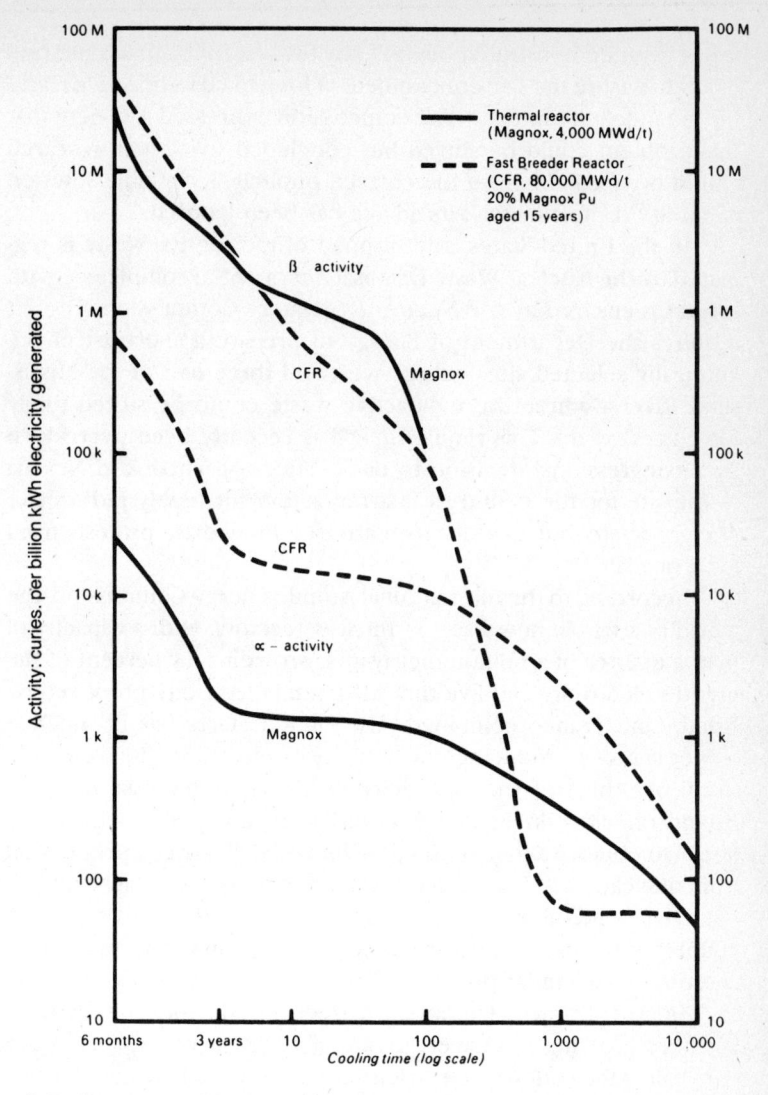

28. Radioactivity of highly radioactive wastes from thermal reactors and fast breeder reactors, after removal of 99 percent of the uranium and plutonium six months after discharge from the reactor. The curves show how the radioactivities decay with time. Decay of α and β rays is shown on separate curves. Magnox is a gas-cooled thermal reactor, and CFR is an experimental fast breeder reactor in Britain. The scales are logarithmic. SOURCE: *Sixth Report of the Royal Commission for Environmental Pollution, Nuclear Power, and the Environment* (Her Majesty's Stationery Office, Cmd. No. 6618, 1976).

it has been demonstrated beyond reasonable doubt that a method exists to ensure the safe containment of long-lived radioactive waste for the indefinite future." The commission expressed the view that this problem could be solved but concluded that much research would be needed before the correct, publicly acceptable solution is found.[69] Unfortunately, its advice has been ignored.

In the United States, safe disposal of radioactive waste is regulated by the Nuclear Waste Disposal Act of 1982; compliance with the act is ensured by the Nuclear Regulatory Commission. The act requires the Department of Energy to prepare a short list of scientifically selected sites—three west and three east of the Mississippi River—where the radioactive waste could be stored safely for 10,000 years. This requirement has recently been overridden by a congressional decision to name Yucca Mountains in Nevada as the site for the country's first repository for highly radioactive nuclear waste, but that decision aroused immediate protests from local officials.[70]

According to the International Atomic Energy Commission, the world as a whole now has 397 nuclear reactors, with a capacity of over a quarter of a million megawatts, providing 15 percent of the world's electricity and five times the total electricity produced by Britain and France combined. The United States has 88 nuclear power stations producing one-sixth of its electricity; in West Germany one-third of the electricity produced is nuclear, in Great Britain one-fifth, in France two-thirds. By the year 2000 France hopes to obtain 80 to 90 percent of its raised electricity production from nuclear fuels. France decided on this program because it lacks natural gas and oil and has only a little coal; it is the most advanced country in the use of nuclear energy.[71] Japan is in the same position and pursues a similar policy.

Since 1982 net generation of electricity in the United States has been growing at an average rate of 2.6 percent per year. More than half of the country's electricity comes from coal, which remains cheap and plentiful (table 5). The contribution of nuclear energy has been rising at an average rate of 10.0 percent per year, nearly four times faster than the rise in total energy generation,[72] and more than half the new generating capacity added in 1986 was nuclear. By 1990 the nuclear contribution is expected to rise from one-sixth to one-fifth, that of coal to remain as now, and that of oil and gas to decrease slightly, to 15.0 percent.[73] There are no further nuclear power stations on order.

TABLE 5

SOURCES OF ENERGY FOR ELECTRICITY
GENERATED IN THE UNITED STATES

	CONTRIBUTION %	CONTRIBUTION % TO NEW GENERATING CAPACITY, 1986
Coal	55.7	44.0
Nuclear	16.1	52.6
Hydro	11.7 ⎫	
Gas	10.0 ⎬	3.4
Petroleum	5.5 ⎭	
(includes petroleum, coke)		
Other	0.5	
(includes geothermal, wood, wind, and solar power)		
TOTAL	2,487,310 gigawatt hours	11,048 megawatts

SOURCES: U.S. Department of Energy, Energy Information Administration, *Electric Power Annual, 1986,* U.S. Department of Energy, Energy Information Administration, *Annual Energy Outlook, 1984,* DOE/EIA–0383.

THE CHERNOBYL DISASTER

The Chernobyl disaster has shaken public confidence in nuclear energy. How did it happen? Could a similar catastrophe occur in Western Europe or America? What do the pros and cons of different energy sources look like now? What would be the consequences of abandoning nuclear energy?

The Chernobyl disaster was caused by faulty design, and by poor training and crass negligence of personnel who carried out a series of explicitly forbidden operations, culminating in the inactivation of the automatic safety controls which should have shut down the reactor when the turbines lost steam pressure.

In principle, the design of the Russian reactor is simple. It consists of rods of uranium oxide, whose neutrons are slowed down by rods of graphite, which the nuclear reaction heats to 1200°C. The uranium and graphite rods are cooled by water under pressure, which evaporates to steam and drives turbines when the pressure is reduced (figure 29).[74] The intensity of the nuclear reaction is moderated by rods of boron, which are lowered to varying depths between the uranium and graphite rods. The reactor was enclosed

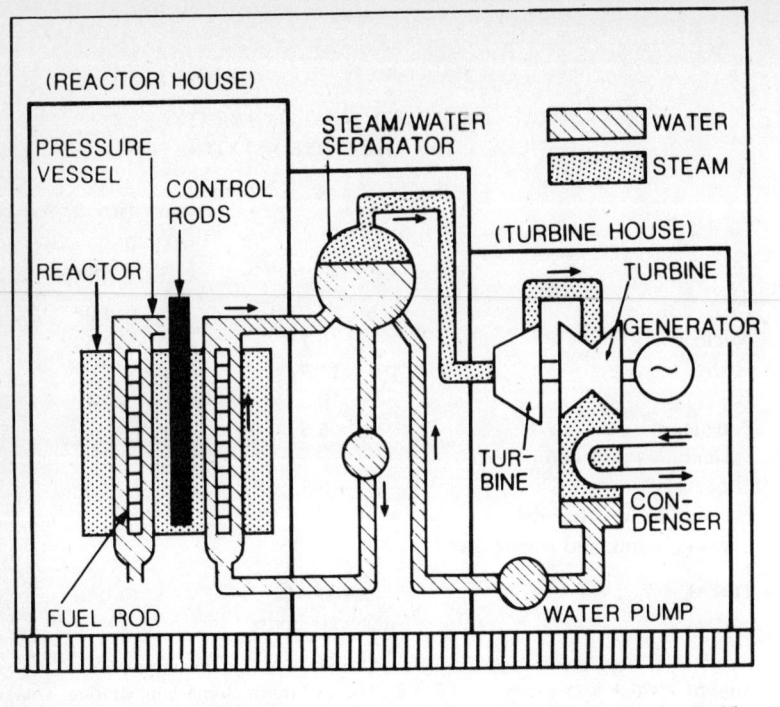

29. The nuclear power plant in Chernobyl. SOURCE: "Shutting the Stable Door," *Nature* 223 (1986): 28.

in a metal housing that was not pressure proof, nor was the building surrounding it.

Some years ago the British government considered the construction of a similar reactor, except that it was to use heavy water rather than graphite to slow the neutrons. When the British engineers examined the Soviet design to see what they might learn from it, they found that it suffered from several weaknesses. The most dangerous one was that accidental formation of steam bubbles in the cooling water raised the energy flux; this would raise the temperature and generate more steam bubbles. This property is known as a positive void coefficient and does not by itself make a reactor unsafe, since a rise in temperature makes the nuclear fuel less efficient and therefore slows the power output. In well-designed reactors this natural decrease ensures safety by compensating for the positive void coefficient.

In the Russian reactor the compensation would occur under normal operating conditions, but when the reactor was run at less than a fifth of its full power, the positive void coefficient became so large that the natural drop in power with rising temperature could no longer compensate for it, and power surged out of control within seconds. Because the reactor had no engineering features to prevent it being run in this dangerous mode, the operators had been given very strict instructions not to do so. At the Vienna meeting of the International Atomic Energy Commission that followed the accident, academician V. A. Legasov said that not even all-powerful Mr. Gorbachev would have been allowed to overrule these instructions, but the operators ignored them anyway, thus causing the accident. He related that the decision not to prevent dangerous operating conditions by built-in safeguards was made in the early 1970s, when the Russian engineers deemed the available equipment to be less reliable than the operators. Legasov admitted that this had been "a colossal psychological error," because no one had foreseen that the operators would become so lax as to lose all sense of danger.[75]

In Western pressure reactors, steam bubbles either reduce the energy flux or raise it so little that the reactors remain under control. If an accident pushes the boron rods out, the reactor compensates for the resulting rise in energy flux and brings it under control. British reactors are designed with fail-safe devices so that any operational error shuts them down. It is not possible to remove all the boron rods at once, as the Russians did before the accident, nor can they be removed rapidly. It must, however, be possible to insert them fast. If the operators mishandle the control rods, the reactor shuts down. The controls of Western reactors indicate any abnormal behavior at once, giving operators time to shut down and cool the reactor, while in the Russian reactor indications of abnormal behavior had to be read on a computer printout. Information given to the operators in the control room is carefully thought out and the reactor designed so that operators do not have to make hurried decisions. Operators are highly qualified and have regular refresher courses that include training on simulators. The Russians admitted in Vienna that they had not realized the importance of such design features and that the training of their operators had been inadequate. Apparently they had no simulators. Finally, Western reactors are surrounded by a concrete housing designed to prevent the accidental escape of radioactivity.

Following a serious accident at the military reactor at Windscale in 1957, Great Britain also set up a Nuclear Safety Inspectorate, which is independent of the nuclear industry and does not license the construction of any power station that does not incorporate these vital safety measures. The inspectorate demands that the probability of a major accident leading to a serious escape of radioactivity be less than 1 in 10 million reactor years. Similar safety standards rule in France, West Germany, Sweden, and Japan.

THE ACCIDENT AT THREE MILE ISLAND

The reactor at Three Mile Island differed in design from that of Chernobyl in having two separate water circulation systems: the primary one, which transferred heat from the reactor to a heat exchanger, and the secondary one, which picked up the heat, generating steam to drive the electric turbines. The accident started when one of the pumps driving that secondary circuit came to a stop. Because of faulty maintenance and bad operating procedures, the spare pump did not automatically take its place. The resulting rise in temperature in the primary water circuit tripped an automatic shutdown of the fission chain reaction. Thus there never was a power surge as there was at Chernobyl. As a result of the pump failure, the water in the primary circuit became too hot and pressure built up, which was relieved automatically by the opening of a valve. Instead of closing again when the pressure was relieved, the valve stuck open. The instruments were designed to signal that it should be closed, but they did not indicate whether it had actually been closed. Through the open valve, weakly radioactive water was released into an isolated storage tank, from which it was then automatically pumped into the open. This was the only radioactivity released by the accident and was later found to have been insignificant.

Nevertheless, events in the reactor now went dangerously wrong. The loss of cooling water from the primary circuit triggered the automatic emergency core-cooling water supply, but the operators did not understand what was happening and switched the water off. A series of further wrong decisions by the operators resulted in the core not being covered by cooling water for about two hours, so that the fuel pellets disintegrated into rubble and fission products were released inside the reactor core. Moreover, the zirconium cladding of the fuel rods reacted with the steam. This reaction produced hydrogen gas, which ignited, but fortunately the fire stopped for lack of oxygen and there was no explosion. The Kemeny Report to the president attributed the accident mainly to poor understanding by the operators and poor management by the utility company, but there were also faults in design. The disassembly of the highly radioactive, broken-up reactor has been an immensely complex task, and is still not completed. The accident has led to important modifications in design and a strict tightening of operating procedures for American nuclear reactors.

In the United States, the Nuclear Regulatory Commission licenses civilian nuclear facilities. Under the Atomic Energy Act, the commission is empowered to see to it "that reactors using nuclear materials are safely designed, constructed and operated to guarantee against hazards to the public from leakage and accident" and that the public is safeguarded "from hazards arising from the storage, handling and transportation of nuclear materials." The commission employs several thousand people and has a budget of over $400 million. It issues licenses for the construction of nuclear power plants; establishes regulations, standards, and guidelines concerning their operation; conducts inspections to ensure compliance with its regulations; and carries out research on safety and environmental questions. It also has an advisory committee of fifteen scientists and engineers, which reviews and makes recommendations on all applications to build or operate nuclear power plants and on related safety matters.[76]

Public fears of nuclear energy have induced several Western European political parties to demand the closure of nuclear power stations. What effects would such a policy have on us and on future generations? This question was recently considered by a committee of the House of Lords.[77] It had fifteen members, among them two biologists, one journalist, one lawyer, one former chairman of the nationalized coal industry, one chemical engineer and former member of the Atomic Energy Authority, and one present director of a West German nuclear energy firm. The other members were laymen with business interests far removed from the nuclear industry. The committee was chaired by Viscount Torrington, also a layman. In fact, there was only one member who had a financial stake in the nuclear industry or was professionally prejudiced in its favor.

The committee listened to witnesses from branches of the nuclear industry from Great Britain, West Germany, France, and Sweden and also to Britain's chief opponents of nuclear energy, the Friends of the Earth. This group's attitude toward the nuclear industry is summarized by one of the witnesses' statements: "If we don't believe in its safety, we cannot use it, whatever the economic consequences. . . . If it were to be believed that one such disaster [like Chernobyl] could be expected somewhere in the world every ten years, or even once a century, there would be few who would contemplate the prospect with equanimity."[78]

The Swedish government has promised to do away with nuclear electricity generation, even though half of Sweden's electricity is now nuclear. When the House of Lords Committee asked the Swedish representative what is to replace that half, he had no answer; the Swedish public opposes any addition to the number of coal-fired power stations because they would aggravate the pollution of lakes by acid rain, and Parliament has virtually ruled out the building of further dams for the generation of hydroelectric power. Nor does Sweden have enough sunlight for significant solar power generation. Some energy can certainly be saved, but not 50 percent. It only remains for the standard of living to be lowered, but this would cause severe unemployment and would be politically unacceptable. I suspect that the Swedish government will not be able to live up to its promise.

The House of Lords Committee concluded, "A decision on the future of the nuclear industry is not simply a decision for or against nuclear power. A decision against nuclear power inevitably means a decision for some other means of electricity generation. Neither the United Kingdom, nor the European Community, nor the world at large, is simply going to do without the electricity that would otherwise have come from nuclear sources. The decision therefore involves a balancing of the advantages and disadvantages of different sources."[79] The public is so used to the comfortable life provided by cheap electricity that people can hardly imagine how tough it can be without it. For example, during the winter of 1987 energy in Romania was so short that the government forbade the heating of offices and apartments above 12°C. This must have cost many old people their lives.

SAFETY OF NUCLEAR POWER STATIONS

The House of Lords Committee believes that application of the highest safety standards to the design and construction of nuclear power plants and to the training of their personnel should make it possible to build and run such plants in a way that reduces the possibility of a catastrophic accident practically to zero. Even though safety standards are very high already, every possible effort should be made to raise them higher still; when faced with having to decide between lower cost or greater safety, safety must always be chosen.

Safety of nuclear power stations has become an international problem. The nuclear industry will flourish only if safety standards everywhere are raised to the highest possible level. The committee therefore favors an international agreement on all aspects of reactor safety: planning, construction, running, training, and safety regulations. To be effective, observance of such an agreement would have to be monitored by an international inspectorate that is at least as powerful as that of the International Atomic Energy Commission, which monitors observance of the Nuclear Nonproliferation Treaty.

I have the impression that Western Europeans worry too much about the safety of their own nuclear power plants; the personnel of these plants would avoid any infringements of safety regulations, if only because such infringements would immediately become public. In Great Britain, the commercially and politically independent Safety Inspectorate ensures strict observance of these regulations. In contrast, the history of the Chernobyl catastrophe shows the dangers that the nuclear industry presents in totalitarian societies, where criticism is muzzled and abuses are kept secret. There has also been doubt about whether all the private utility companies operating nuclear plants in the United States have the financial and technical means of ensuring maximum safety.

Even if the possibility of another Chernobyl can be excluded, the public is concerned that radioactivity escaping from nuclear power plants operating normally might raise the incidence of cancer and hereditary diseases. But these fears are unfounded because the radioactivity released by nuclear power plants constitutes no more than a tiny fraction of the natural radioactivity to which all of us are exposed. Ordinary soil contains surprisingly large quantities of radioactive elements: 1 cubic meter of average English garden soil contains 17 kilograms of potassium, 2 grams of which are radioactive, in addition to 15 grams of thorium and 5 grams of uranium. Coal also contains uranium; thus, coal-fired power stations spread annually about 120 tons of uranium in their ash and in the atmosphere. Figure 30 shows the radiation impinging upon us from different sources. Air travel at high altitudes and all industry, including nuclear power plants, contribute no more than 1.5 percent, while the remaining 98.5 percent arises from unavoidable natural sources. Today about 50,000 out of every 1 million deaths are caused by cancer. About 200 of these probably result from radiation, and only 3 of these are caused by air travel and industry. In general, the cancer risk resulting from all the radiation to which we are

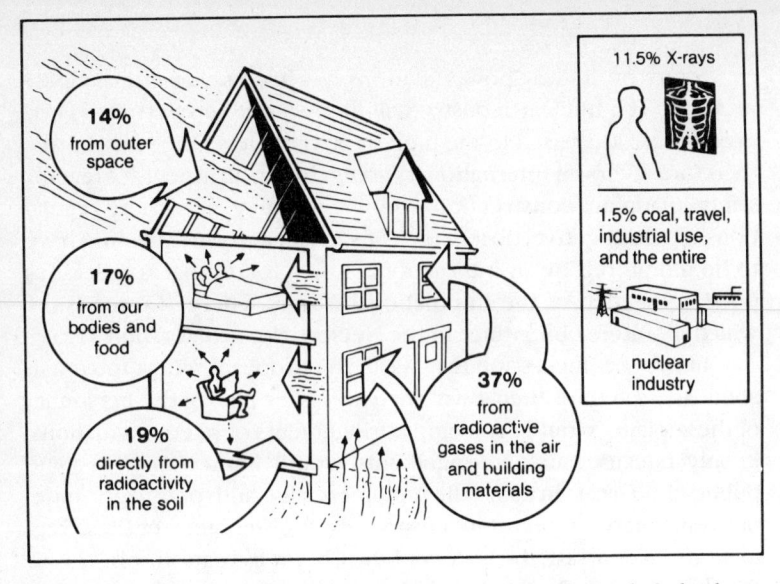

30. Sources of radiation to which we are constantly exposed. Only about one-seventieth comes from industry, including the nuclear industry. SOURCE: Walter Marshall, "Tizard Lecture," *Atom*, June 1986, 1–8.

exposed daily is equivalent to no more than that of regularly inhaling five puffs of a cigarette once a week.[80] However, in homes that rest on highly radioactive rocks, the cancer risk from background radiation can be much higher, as was recently found in certain areas of the United States. Apparently, hundreds of thousands of Americans living above such rocks regularly receive annual doses of radiation as high as the people who lived near the Chernobyl reactor after its explosion.[81]

The British Office of Population Censuses and Surveys recently published a report titled "Cancer Incidence and Mortality in the Vicinity of Nuclear Installations in England and Wales, 1959–1980," which the epidemiologist Sir Richard Doll and his colleagues at Oxford subjected to a statistical analysis. They have summarized their results as follows:

> These data show conclusively that there has been no general increase in cancer mortality in the vicinity of nuclear installations in a 22-year period beginning several years after the opening of the installations that have released the largest amounts of radionuclides to the environment. On the contrary, the mortality from cancer has tended to be lower in the Local Authority

Areas in the vicinity of nuclear installations than in the control Local Authority Areas selected for their presumed comparability with the former. This is unlikely to be due to a protective effect of ionizing radiation and suggests that, despite the efforts that were made to choose comparable control areas, there were non-installation differences between the populations relevant to the risk of dying from one or other type of cancer.

Detailed examination of the few types of cancer that were relatively more common in the installation areas suggests that several of the differences were most likely to be due to chance, diagnostic artefacts or social factors rather than to any hazard specifically related to the installations. One disease provides a possible exception: namely, leukaemia in the age group 0–24 years. Two other diseases need further investigation, multiple myeloma and Hodgkin's disease in the older age group 25–74 years. The excess mortality rates recorded from these cancers were not large, and it has yet to be established that they are not due to a general confounding by other environmental or socio-economic factors.[82]

The increase in leukemia in the age group 0–24 arises from a fourfold rise in deaths from leukemia of children under age ten who lived close to nuclear plants built before 1955, but this may be accidental, because the geographic incidence of leukemia is uneven and the causes of other clusters are unknown. There has been no increase in leukemia among children living close to plants that have been built since, though there has been a recent report of six cases of child leukemia near a nuclear reprocessing plant in Scotland which are as yet unexplained.

ALTERNATIVES TO NUCLEAR ENERGY

Could the risks of nuclear energy be at least reduced by exploiting alternative energy sources, even if their supply is limited? The most obvious alternative is coal, of which the world contains enormous reserves and which can be converted to either gaseous or liquid fuel. When coal was burned without special precautions as the main source of energy, pollution by smoke in Manchester early in this century became so bad that over half the children had rickets because they rarely saw the sun and 1 kilogram of dust fell on every square meter in one year. The air in Britain is much clearer now, partly because coal consumption has fallen to less than half

31. Penetration of sulfuric and hydrochloric acids into the Verona stone has ruined this statue outside a church in Venice. The white stains are pigeon droppings. SOURCE: Guido Biscontin and Luigi Cattalini, "Venice Regained," *Chemistry in Britain* 16 (1980): 360.

what it was in 1910, and partly because modern methods of combustion have reduced the emission of solid particles in coal smoke to perhaps one-thousandth of what it used to be. The most noxious of the gases in coal and oil smoke is sulfur dioxide. The wind now carries sulfur dioxide from British and Continental smokestacks to Scandinavia, where it is killing the fish in lakes and rivers. In Venice sulfur dioxide generated by industry in neighboring Mestre and by motor traffic on the canals makes the beautiful facades of churches and palaces crumble (figure 31).[83] In West Germany sulfur dioxide and nitric oxides may be killing the forests. Increased burning of any fossil fuel may worsen pollution by sulfur dioxide and nitric oxides unless most of these substances are removed from industrial smoke and car exhausts. Their removal from car exhausts will add to the cost of driving; in the United States, where coal is cheap, the cost of their removal from the smoke of coal-fired power stations can be of the same order as the cost of the coal itself. Governments may therefore be reluctant to impose such restrictions, but even so, this reluctance may be shortsighted.

Coal mining is one of the riskiest occupations. Out of 232,000 miners employed in Britain in 1978–79, 72 were killed and 480 seriously injured, an accident rate that is about ten times higher than the average in manufacturing industry. In addition, pneumoconiosis (a lung disease caused by coal dust) contributed to the deaths of over 200 miners, but these were all men employed before 1955. Since then, efficient suppression of dust has almost entirely prevented that disease.[84]

By contrast, nuclear energy causes no pollution other than an insignificant heating of the ocean as long as the radioactive waste is safely contained. In fact, mankind's entire energy consumption heats the Earth directly only very little, but the carbon dioxide set free by the burning of fuels threatens to heat it indirectly. Since 1860, when it was first measured, the carbon dioxide content of the atmosphere has risen by 19 percent (figure 32). The consumption of fossil fuels has recently been rising by 2 percent per year. If this rate of increase were to remain the same until 2050 and consumption held constant thereafter, the carbon dioxide content of the atmosphere by 2065 might be double that of 1860. Were fuel consumption to rise by 3 percent until 2025 and to hold constant thereafter, the doubling might be reached by 2040.[85] What effects would this have?

Carbon dioxide absorbs heat emitted by land and water and radiates it back to Earth. Computer models of the atmosphere indicate that a doubling of the carbon dioxide content would raise the average temperature on Earth by 1.5 to 3.0°C (the most probable rise is 2.8°C), but this increase would be distributed unevenly, with the smallest rise (2.0°C) in the tropics and the largest (9.0°C) at high latitudes, where it would extend the cultivable season and improve the water supply. At moderate latitudes, however, reduced precipitation and increased temperatures would lead to droughts. Given sufficient irrigation, increased atmospheric carbon dioxide would promote plant growth, because it accelerates photosynthesis and reduces water loss. In the Arctic, the Northwest and Northeast passages might become navigable.

All this suggests that mankind might adapt itself to the climatic changes and that some countries would even profit from them. There is, however, one great danger. At present the sea level is rising annually by 1.5 millimeters, partly because of the increased volume of carbon dioxide absorbed by the sea and partly because of increased melting of the arctic and antarctic ice. Were a large part of the antarctic ice that is now submerged under the sea to

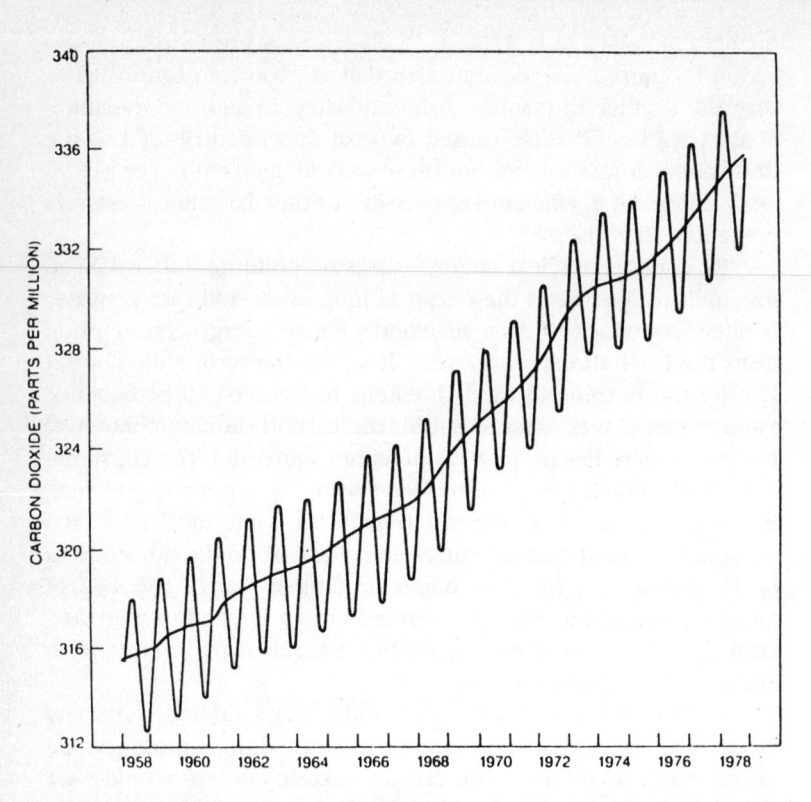

32. Parts per million of atmospheric carbon dioxide on a mountaintop in Hawaii, 1958–1978. The seasonal fluctuations result from increased absorption of carbon dioxide by plants in the spring. SOURCE: Roger Revelle, "Carbon Dioxide and World Climate," *Scientific American* 247 (August 1982): 33–41.

break up and melt, the sea level would rise much faster, until it reached a level 5 to 6 meters above its present one. Many countries now exhibit rock terraces 5 meters above sea level with fossils showing that they were once submerged. The terraces are 125,000 years old, corresponding to the warmest period between two ice ages (figure 33). It would take perhaps as long as two to five hundred years for this to happen again, but the sea would rise by 2 to 3 centimeters a year, enough to soon submerge not just Venice but also New Orleans, London, Hamburg, the Netherlands, the plane of the Po, and many other fertile and densely populated parts of the world.[86] This is a risk we cannot afford. Yet removal of carbon

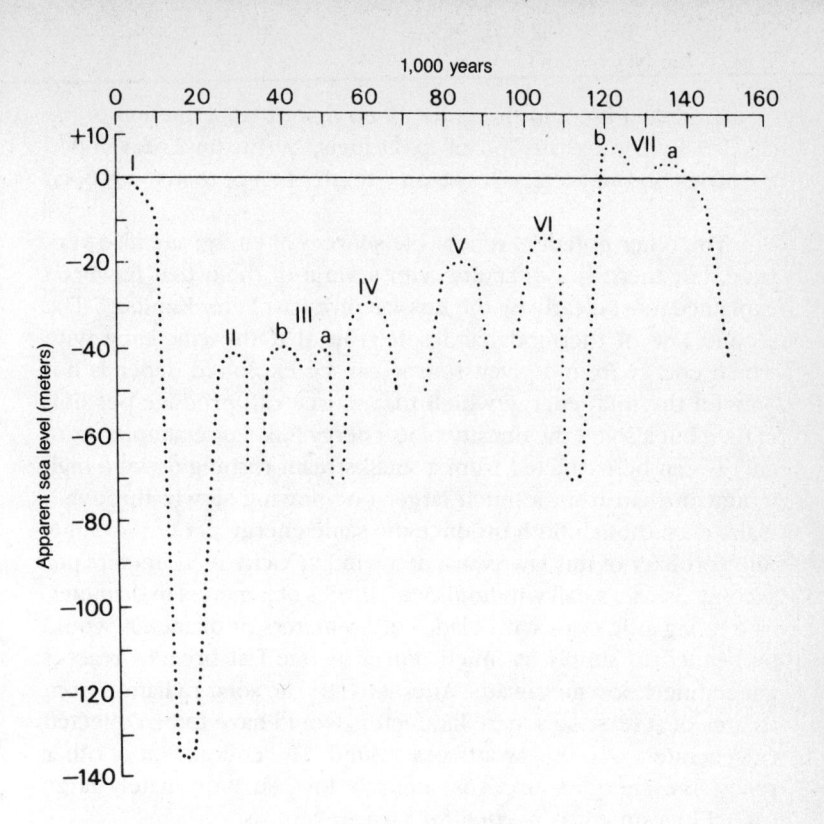

33. Maximal sea levels in the past 150,000 years. The terrace 5 meters above the present sea level 120,000–125,000 years ago was formed after the breakup and melting of the west Antarctic ice cap. I–VII represent warm periods between ice ages. SOURCE: W. S. Moore, "Late Pleistocene Sea Level History," in *Uranium Series Disequilibrium: Application to Environmental Problems,* eds. M. Ivanovich and R. S. Harmon (Oxford: Clarendon Press, 1982).

dioxide from smoke would cost as much energy as the burning of the fuel had produced.

Could the expected energy gap be filled from renewable energy sources, such as hydroelectric power, biological fuels, heat inside the Earth, sunshine, tides, and wind? The use of hydroelectric power has been growing worldwide at an average annual rate of over 5 percent for the last fifty years, but even so it now supplies no more than 7 percent of the world's energy. Hydroelectric power generation is clean, prevents flooding of rivers, and can help irrigation of land, but it is expensive to start with and can spoil the countryside. In developed countries much of its potential is already being exploited, and it could therefore fill no more than a fraction of the gap left by the exhaustion of fossil fuels, but in underdeveloped countries on the average less than a tenth of its potential has been tapped.

The other potential renewable sources of energy are also very large, but there is a difficulty with several of them that has been explained most clearly by the Russian physicist Peter Kapitsa.[87] The second law of thermodynamics tells us that the efficiency with which energy from a given source can be exploited depends not only on the total energy which that source can produce per unit of time but also on the density of its energy flux. For example, more energy can be extracted from a small stream rushing down a high mountain than from a much larger one flowing slowly through a plain, even though both produce the same energy per unit of time. The corollary of this law is that at a wind velocity of 10 meters per second, 25,000 small windmills with blades of 8 meters in diameter, or 250 gigantic ones with blades of 80 meters in diameter, would be needed to supply as much power as one fast breeder reactor generating 1,000 megawatts. Alternatively, the solar radiation from an area of at least 50 square kilometers would have to be collected to generate 1,000 megawatts year round. The energy flux of other renewable energy sources is similarly low, so inordinately large capital investment is needed for meager returns.

The British Department of Energy estimates that by the turn of the century all renewable sources of energy combined could supply only about 10 percent of Britain's total energy demand, although others consider this estimate too low. In the United States, President Carter gave 20 percent as the goal; this includes additional hydroelectric power, but experts regard 12 percent as a more realistic figure. By contrast, renewable sources might make substantial contributions in southern countries. Where each village needs to generate no more than 50 kilowatts for irrigation and lighting, solar and wind energy or methane generated by fermentation of cow dung would be more economical than energy from a large, distant power station. Brazil hopes that by 1985, 20 percent of all its liquid fuel will come from fermentation of cane sugar and cassava.

Finally, there is the distant hope of energy from controlled nuclear fusion.[88] This process could not be abused by terrorists. It would be inexhaustible and would therefore provide us with enough energy indefinitely, but the technical problems with nuclear fusion are formidable, and it is not yet clear when, if ever, they will be solved. Nor is it clear that nuclear fusion would not also pose problems of radioactive pollution.

What would happen if the richer nations decided not to build any more nuclear power stations and to close down the existing ones because the risks are too great? Sir Hermann Bondi, former chief scientist at the British Department of Energy, has pointed out to me that such a decision would exacerbate the present inequality of energy distribution because it would drive the price of fossil fuels up to a level that would put them quite out of reach of the poorer countries. This is already happening in India, which now spends 70 percent of its export revenue on oil. Hence, it is vital, Bondi argues, that scientists find technical solutions for the hazards of nuclear energy and also convince the public of its safety as an alternative to fossil fuels. Otherwise, shortage of these fuels may bring much of agriculture and industry to a standstill in countries that can no longer afford them. The same view has been expressed by the International Atomic Energy Commission.

The late Sir Martin Ryle, a Nobel laureate in physics and the most outspoken critic of official British energy policy, took a totally different view. He argued that even in the most optimistic forecasts, by the end of the century nuclear energy will be able to provide no more than a small fraction of the total energy that is now supplied from fossil fuels. He estimated this total in Britain as 320,000 megawatts, of which only 45,000 would come from nuclear power. To scale up present electricity generation to anywhere near this total would be impracticable.

He argued, therefore, that the large energy gap can be filled only by reducing waste and by using renewable energy sources to the full. He calculated that if the capital cost of a nuclear power station were spent instead on energy-saving devices, some three times more energy could be saved than the station would produce in its lifetime. This is particularly true in the United States, the country most profligate in its waste of energy. The European Economic Commission also regards energy saving as equivalent to the best sources of additional energy. Ryle considered that the efficiency of windmills, solar heat collectors, and heat storage devices is improving so fast that the official estimates of their possible contributions are too low, and he reported that in Denmark a figure of 44 percent has been quoted as the contribution of renewable energy sources by the year 2000, and in Canada 100 percent by 2025. But others are skeptical of these estimates for the reasons already given.

Ryle regarded the generation of nuclear power as not worth the risk.[89]

I believe that the possibility of another severe nuclear accident or of nuclear terrorism should be weighed against the more likely risk of an acute world energy shortage. I have the impression that the opponents of nuclear power do not give enough thought to the unemployment, the famines, the social unrest, and the international tensions that an energy shortage would generate. For example, the economic recession that began in 1973 was caused by the balance of payments problems resulting from inflated oil prices.

That rise in price was the result of a political decision by the oil-producing countries, made while oil was still plentiful. When it becomes genuinely short and harder to extract, its price is likely to rise steeply to much higher levels than in the 1970s. Figure 25 shows that this is not something we should forget, or presume might happen one day in the unforeseeable future, but something that will hit our children and grandchildren. Shortage of energy will exacerbate the plight of the poorest and most overpopulated countries and will force even the richest to lower their standard of living. We have seen that most of the world's coal reserves are located in the Soviet Union and the United States; their military strength could tempt them both to monopolize most of the remaining sources of oil and natural gas, which would in turn allow them to delay that painful lowering of the standard of living in their own countries while weaker nations may lack the energy they need to survive. Exploitation of nuclear energy would make the world's resources of fossil fuels last longer and reduce the weaker countries' dependence on them. People who clamor for the abandonment of nuclear energy should weigh the avoidable risks of nuclear accidents against the certainty of a future energy shortage and all the suffering it is likely to entail. They should also remember the appalling threats posed by the greenhouse effect.

There remain some vague hopes of escaping the shortage of energy. The world has large deposits of methane hydrate, a compound consisting of methane molecules enclosed in cages of water molecules that outnumber the methanes in a ratio of either 6 or 17 to 1. The deposits lie under the permafrost in Siberia and Alaska and also in deep ocean sediments. Permafrost extends to a depth of 300 to 600 meters, and the methane deposits vary in thickness between 300 and 1,000 meters. Deposits of methane hydrate may well exceed the known reserves of conventional natural gas, but it is not clear if methane hydrate can be extracted at a lesser cost in energy than that released on burning the methane.[90]

The British astrophysicist Thomas Gold has predicted the presence of huge reserves of methane in deep layers of the Earth's crust.[91] In Sweden this hypothesis is now being put to the test by the drilling of a many-kilometers-deep hole. Gold's ingenious and bold hypotheses have often proved true. If he turns out to be right this time, mankind would be well supplied with energy into the distant future, but the world would still be faced with the ever rising carbon dioxide content of the atmosphere. Only nuclear energy or large-scale conversion of solar energy could prevent this from happening.

ENERGY CONSUMPTION, FOOD SUPPLY, AND POPULATION GROWTH

North America and Western Europe consume 55 percent of the energy used in the world, and they want more each year to sustain further growth of an already absurdly high standard of living. The average Westerner spends more on tobacco, drink, and cosmetics in a year than the average Indian's annual income. In many developing countries food production cannot be raised above subsistence level for lack of energy and capital to produce nitrogenous fertilizers. What is needed, therefore, is not a rise in energy consumption in the richest countries but a more equitable distribution of the energy that is produced in the world. Figures 25 and 26 show the stark contrast between accelerating consumption and shrinking supplies of fossil fuels. At present it does not look as though enough energy will be available to sustain indefinite economic growth; in the next century all the resources of science and technology—including energy conservation, power from renewable energy sources, and nuclear power—will be needed to ensure the survival of civilized life, especially in the poorer countries.

The world population now doubles every twenty-five to thirty years, and it is expected to reach about 6 billion by the year 2000 (figures 34 and 35), by which time the cultivated area per head of population will have shrunk to one-eighth of a hectare. Table 6 shows the frighteningly high rates of population growth in many

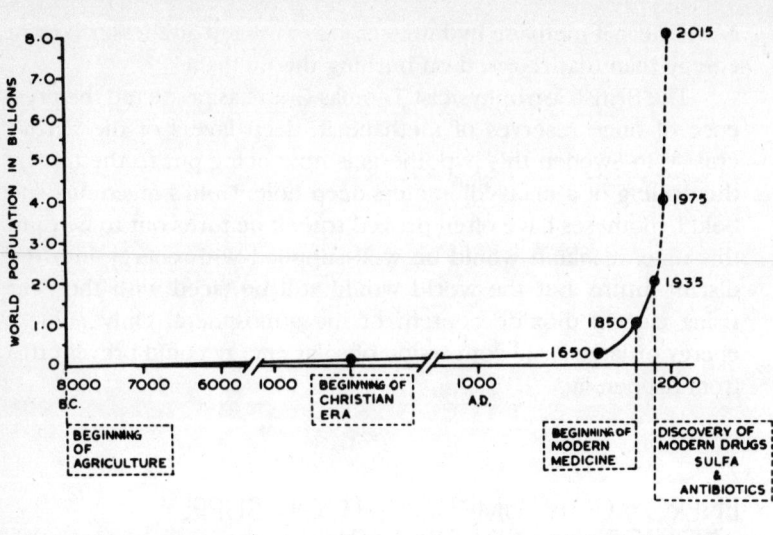

34. World population 8000, B.C.–A.D. 2000. SOURCE: M. S. Swaminathan, *Global Aspects of Food Production* (Geneva: World Meteorological Organisation, World Climate Conference, 1979).

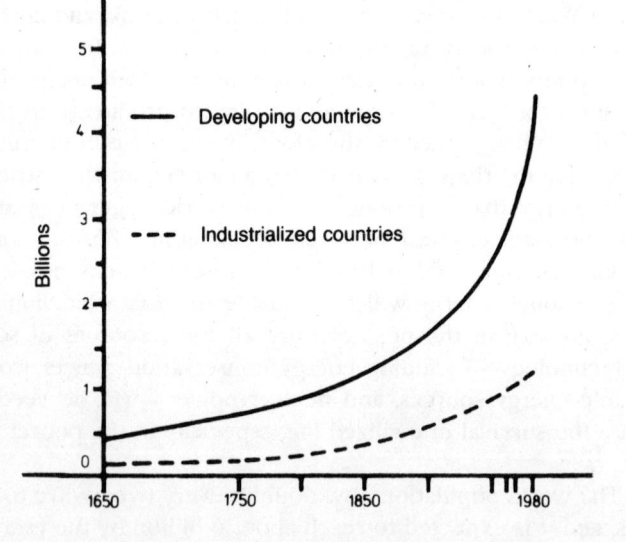

35. Populations in industrialized and developing countries, 1650–1980. SOURCE: M. S. Swaminathan, *Global Aspects of Food Production* (Geneva: World Meteorological Organisation, World Climate Conference, 1979).

TABLE 6

VITAL STATISTICS OF TYPICAL DEVELOPING AND DEVELOPED COUNTRIES

COUNTRY	POPULATION GROWTH, % PER YEAR, 1973–1984	% MARRIED WOMEN USING CONTRACEPTION, 1983	INFANT MORTALITY UNDER 1 YEAR (per 1,000) 1965	INFANT MORTALITY UNDER 1 YEAR (per 1,000) 1984	% CHANGE IN BIRTHRATE, 1965–1984	POPULATION (MILLIONS) 1984	POPULATION (MILLIONS) 2000	GNP IN $ PER CAPITA 1984	% CHANGE IN GNP PER CAPITA, 1965–1985
Ivory Coast	4.5	?	176	106	+ 2.4	10	17	550	+ 0.2
Kenya	4.0	17	113	92	+ 4.3	20	35	275	+ 2.1
Zaire	3.0	3	142	103	− 28.3	30	47	125	− 1.6
Nigeria	2.8	5	179	110	− 3.4	96	163	660	+ 2.8
Egypt	2.6	30	173	94	− 17.2	46	65	650	+ 4.3
Brazil	2.3	50	104	68	− 24.6	133	179	1,550	+ 4.6
India	2.3	35	151	90	− 27.1	749	994	235	+ 1.6
China	1.4	71	90	36	− 51.3	1,029	1,245	280	+ 4.5
USA	1.0	76	25	11	− 19.1	237	263	13,900	+ 1.7
USSR	0.9	?	30	NA	+ 8.9	275	307	NA	NA
Italy	0.3	78	38	12	− 46.1	57	59	5,800	+ 2.7
Great Britain	0	77	20	10	− 28.0	56	58	7,750	+ 1.6
West Germany	− 0.1	NA	26	10	− 46.3	61	60	10,000	+ 2.7

SOURCE: *World Development Report, 1986* (Oxford: Oxford University Press for the World Bank, 1986).

developing countries. Growth rates of 2.3 to 4.5 percent correspond to doubling times of only twenty-eight to fifteen years (figure 36).[92] By the year 2000 the populations of many already very poor countries—such as Kenya, Zaire, Nigeria, and Egypt—will have reached enormous numbers. In Nigeria alone, the additional population will be over 50 million, nearly as many as the entire present population of West Germany, while the present income per head is only one-fifteenth of the average German income. The projected increases result partly from lower infant mortality and increased life expectancy (figure 37). Birthrates in developing countries have actually dropped in the past twenty years, a trend that has raised hopes that the world's population may stabilize eventually at about 10 billion, perhaps because this is such a nice round number. Despite the successes of scientific agriculture, the world's food production is unlikely to keep pace with the rise in population. There are several reasons why the outlook is grim. One is the conversion of some of the best farmland to the building of houses and roads. In the United States about 400,000 hectares of prime cropland were converted to nonfarm use each year between 1967 and 1977, amounting to a loss of nearly 3 percent in ten years. In West Germany the loss between 1960 and 1970 amounted to 2.5 percent; in Britain and France to 2 percent. From 1980 to 2000 the area of land in the world that is under cultivation with cereals is expected to rise by only 9 percent, while the world population is likely to rise by 40 percent, which will reduce the cultivated area per person from 0.17 to 0.13 hectares (table 7).[93] In addition, the land being brought into cultivation in Third World countries is generally of marginal fertility.

TABLE 7

WORLD POPULATION AND AREA IN CEREALS, 1950–2000

	POPULATION (BILLIONS)	AREA IN CEREALS (MILLION HECTARES)	AREA PER PERSON (HECTARES)
1950	2.51	601	0.24
1980	4.42	758	0.17
2000	6.20	828	0.13

SOURCE: *World Development Report, 1986* (Oxford: Oxford University Press for the World Bank, 1986).

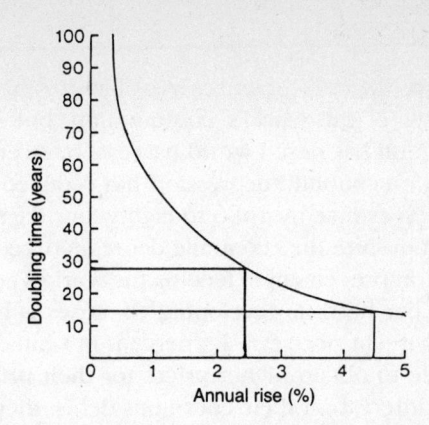

36. Mathematical relationship between percentage growth and doubling times.

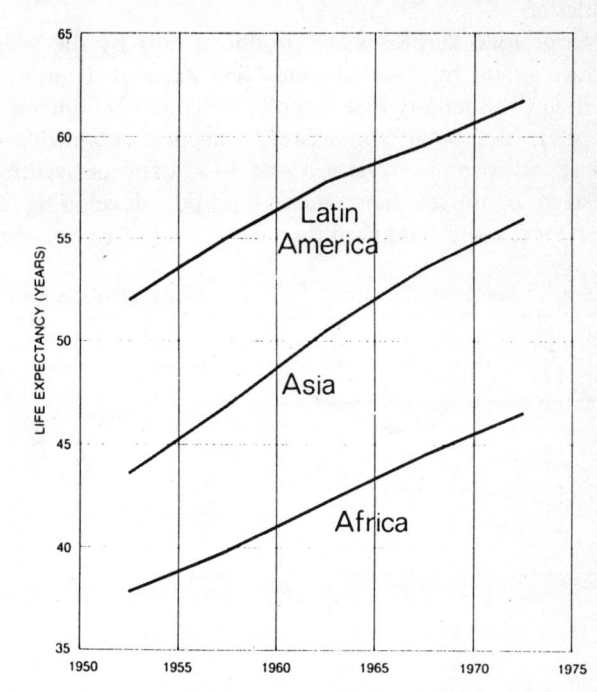

37. Life expectancies in Africa, Asia, and Latin America, 1950–1975. Average life expectancy is rising equally fast in different developing countries, but it remains markedly different on different continents. In the early 1970s it amounted to about forty-seven years in Africa, fifty-seven in Asia, and sixty-two in Latin America. D. R. Gwadkin and S. K. Brandel, "Life Expectancy and Population Growth in the Third World," *Scientific American* 246 (May 1982): 33.

In 1980 the world's grain reserves had shrunk alarmingly, to only forty days of the world's consumption, but since then the Green Revolution has raised world harvests by over 2 percent per year, while the economic depression has reduced demand. As a result the reserves rose by 1986 to eighty-four days of world consumption. Yet despite the economic depression, recent years have brought some improvement in feeding the world's population: food consumption per head in developing countries is rising by an average of 3.0 percent per year, 4.5 percent in China. If these countries were able to obtain higher prices for their primary products or pay lower interest on their enormous debts, they could readily buy and consume the grain surplus of the Western democracies. This surplus will be needed to keep pace with the rising world population.

Large food surpluses are produced only by the Western democracies, and by Australia and New Zealand. Figure 38 shows that their productivity rose steeply; developing countries achieved a gentler rise, while the Eastern European countries failed to achieve any rise at all. As a result, food exports by the Western industrial countries have increased, but developing countries now import more food than they export, and East European coun-

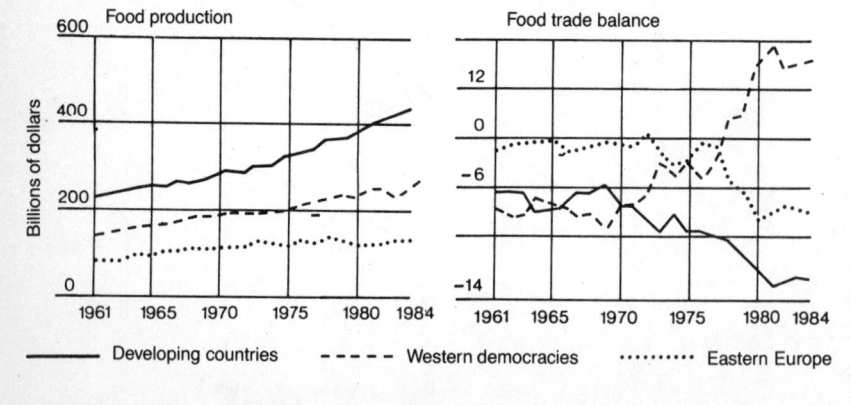

38. Production and trade balance of food in billions of dollars in developing countries, Western democracies, and Eastern Europe, 1961–1984. Note the contrast between the steep rise in Western democracies and the stagnation in Eastern Europe. There is growing concern that the Western industrial countries are the only net food exporters now left. SOURCE: *World Development Report, 1986* (Oxford: Oxford University Press for the World Bank. 1986).

tries, who own some of the world's richest farmlands, import even more.

It looks as though mankind is eating away the agricultural resources on which its children and grandchildren will have to live. Hunger for food or for profits or, in the Communist states of Eastern Europe and the Soviet Union, bad planning and mismanagement, makes farmers exploit their soil beyond its capacity. This improvidence is leading to a gradual reduction in the soil's fertility, while the fertility of man continues to rise. Third World politicians blame their people's poverty and malnutrition on exploitation by the West. There is substance in many of their accusations, but I doubt that even the most enlightened and generous policies of the West will be able to prevent catastrophic famines unless Third World countries themselves take action to conserve their soil and slow down their population increase. Still, as John Steinbeck said, "You can't forbid people being born, at least not yet."

It is often said that contraception is employed only by populations that have already attained a high standard of living, and that it is not possible to persuade uneducated people or people of non-European culture to use birth control. China has now shown that it is possible to explain, even to illiterate people, that the limitation to one child per couple is the only way of securing children's future. China is also producing contraceptives cheaply on a large enough scale, and in a socially acceptable form, for its population of more than 1 billion people.[94] Tragically, Muslim societies are said to resist any form of birth control. In consequence, the populations of Egypt and of several other Muslim countries are now rising at a doubling rate of only twenty years. I can see no way of providing these multitudes with the gainful employment and the food needed to keep them alive. In the field of contraceptives, the balance between benefits and risks lies overwhelmingly on the side of the benefits. Even in Britain, with its excellent maternity services, the health risks of oral contraceptives are much less than those of childbirth. The religious and social prejudices that are retarding the use of contraceptives in many countries, and especially the opposition to their introduction in Third World countries by the Catholic church, will exact a heavy toll in human suffering. Among developed countries the percentage of women using contraception is highest in predominantly Catholic Italy, France, and Belgium, despite the Catholic church's opposition. It is also increasing slowly in the developing countries (table 6).

SCIENCE AND POLITICS

Can science do anything to lessen either international or national tensions? Science is just knowledge and has no political content, but Karl Popper, a philosopher who has given much thought to the methods of science and their application to society, has shown that science can make at least a modest contribution by guiding people toward a scientific attitude to political problems. In a study of the history of political science, called *The Open Society and Its Enemies*, he attacked those philosophies that later became recipes for tyranny.[95] All such philosophies have invoked laws according to which human society is bound to develop along certain predetermined paths. Plato pictured that path as a degeneration from the ideally perfect state and suggested that this degeneration must be arrested by the authoritarian rule of the wise few over the stupid multitude. George Orwell's *Nineteen Eighty-four* is a brilliant caricature of Plato's *Republic*. According to Marx's historical laws, industrialization is bound to lead first to class war, then to the victorious revolution and dictatorship of the proletariat, and finally to the withering away of the State.

Popper started from the premise that society is too complex a fabric to derive from its past laws on which to base prophecies about its future. Anthropological research has shown that, contrary to Plato, primitive societies are far from being models of moral perfection, and their social systems have generally remained static. Contrary to Marx, revolutions have occurred only in agrarian societies subjected to feudalism, while in highly industrialized societies the working class has tended to merge with the bourgeoisie. In fact, all so-called laws of history enunciated in the past have been falsified by events. Popper argued that the future depends only on ourselves. Hence there is no law that makes either international war or national class war inevitable. It is up to us to see that they do not happen.

Lord Acton has said that democracy consists of forestalling revolution by timely reform. Popper argued that such reforms should be undertaken in the spirit of science, in which knowledge is only provisional and natural laws are regarded as hypotheses conceived to be tested experimentally. No hypothesis can ever be fully proved, because there might always be an experiment that would falsify it. Defined like this, Galileo's law of gravitation is scientific, but Freud's axioms of psychoanalysis are not. Experiments may force us to modify our original hypothesis so that gradually it approaches nearer and nearer to the truth. As Peter Medawar put it, "Scientific reasoning is a kind of dialogue between the possible and the actual, between what might be and what is in fact the case."[96] Popper argued that political, economic, and social problems should be approached in the same pragmatic manner rather than with dogma. Since human society is exceedingly complex, the outcome of even the best-planned reforms will always be uncertain. Only gradual and nonviolent changes have any likelihood of producing the results desired, and even then they may be accompanied by adverse and unforeseen side effects, rather like those of a new drug. Popper suggested that the social sciences should learn to predict such side effects, and that politicians and administrators should continually modify their policies in light of them. Such an open-minded approach is possible only in the atmosphere of free discussion of a democracy.

In Italy, many young people say, like Mourlan in Roger Martin Du Gard's *Les Thibault*, "Everything has got to be smashed to start with. Our whole damned civilization has got to go before we can bring any decency into the world." A young Italian scientist living comfortably in Switzerland recently said to me, "Italian universities are so rotten that you have to wait for their collapse before you can start building decent ones." He should have known better.

Popper showed that such a "clean slate" policy is unlikely to produce the desired result, because destruction of the social fabric also destroys the institutions and moral values, the very decency on which a new improved society could be built. Popper's arguments have been confirmed by the disastrous results that followed when such a policy was actually carried out by the Communist fanatics in Kampuchea. It led to the killing of 2 million people, to famine, disease, and invasion.

Science is the triumph of reason. Bertrand Russell said, "Rationality, in the sense of a universal and impersonal standard of truth, is of supreme importance, not only in ages in which it easily prevails, but also, and even more, in the less fortunate times in which it is despised and rejected as the vain dream of men who lack the virility to kill when they cannot agree."

However, reason is not enough. The twelfth-century French philosopher Peter Abelard wrote that "science without conscience is the death of the soul." Not just of the soul. In the 1930s German geneticists, anthropologists, and psychiatrists decided that the German race must be cleansed of "inferior" types whose conditions they regarded as hereditary. These included inborn subnormals, schizophrenics, manic-depressives (like Virginia Woolf), inborn epileptics (like Fyodor Dostoyevsky), and severe alcoholics (like Ernest Hemingway). At first such people were sterilized, and by the time of Germany's attack on Poland, some 350,000 to 400,000 sterilizations had been done. After the outbreak of war, it was decided to kill them instead. Psychiatrists and other doctors selected those to be sent to special killing stations. The exterminations were planned and organized by a staff of about three hundred at the Institute for Neurology of the Charite Hospital in Berlin. The building was destroyed later in the war. Some now propose the erection of a memorial stone where the hospital once stood with the following inscription:

In Memory of the Policy of Extermination,
and
In Honor of Its Forgotten Victims

Here, in Tiergartenstrasse 4, the National Socialists organized the first mass murders. More than 200,000 defenseless people died by gas, sleeping drafts, or planned starvation. Their lives were considered worthless, their murder was called euthanasia. The perpetrators were scientists, medical doctors, nurses, and officials of the administrations of justice, health, and employment. Their victims were poor, desperate, disturbed, or in need of help. They came from psychiatric clinics, old people's homes, children's and army hospitals, camps, and community homes. The number of victims was large, and only few were convicted of their murder.[97]

Neurologists and others eagerly grasped the "gigantic opportunities" that the bodies of the victims offered for research. Their pickled brains were sent by express post as urgent war materials to the Kaiser Wilhelm Institute for Brain Research in Berlin and other institutions.[98] The results of studies on these organs were published in scientific journals and presented at scientific meetings, whose proceedings offer no evidence of a protest from any of those present. Protests did come, mainly from clergymen, who were often imprisoned for them. Some of the scientists and medical men who carried out this work are still alive, in comfortable retirement, and apparently they look back smugly on their complicity in mass murder,[99] but there is now a growing movement to make the German people aware of these terrible happenings. They prove that reason alone, without compassion and a sense of right and wrong, does not prevent people from committing the most appalling crimes.

After the Second World War, when the United States had the monopoly on the atomic bomb, the Hungarian-born mathematician John von Neumann advocated that the bomb should be used to knock out the Soviet Union before the Russians developed their own atomic bomb. Fortunately, compassion prevailed over this cold reason, which would not have shrunk from killing millions for the crime of living under the wrong regime. More and more politicians have now come to realize that there is no war that would make the world safe for either capitalism or communism, or for any militant creed or race. The dangers of nuclear war are infinitely greater than those of accidents at nuclear power stations. A single thermonuclear bomb could kill several million people, many of them slowly and terribly painfully; it would make 50,000 square kilometers uninhabitable for a month, of which 3,000 would still be uninhabitable after a year.[100] Survivors in shelters would have no place to go that was not lethally radioactive, and nothing to eat or drink that was not radioactively contaminated. The Soviet Union and the United States each have about ten thousand such bombs and are constantly building more, enough to kill most of each other's populations and to make the greater part of the world's cultivable lands infertile and uninhabitable.

Everyone with common sense must help to prevent such a catastrophe. Young people who now demonstrate against nuclear power stations would do better to concentrate on the physicists who keep pressing new weapons on the military, on the industrialists who compete for their manufacture, on the strategists who devise new targets for the weapons, and on the politicians who believe that these weapons will enhance their power.[101]

A nuclear war would destroy everything that has been built up over centuries without giving us any control over what, if anything, will rise from the ashes. We must work for the application of science to peace and a more just distribution of its benefits to mankind.

SCIENCE
IN
WAR

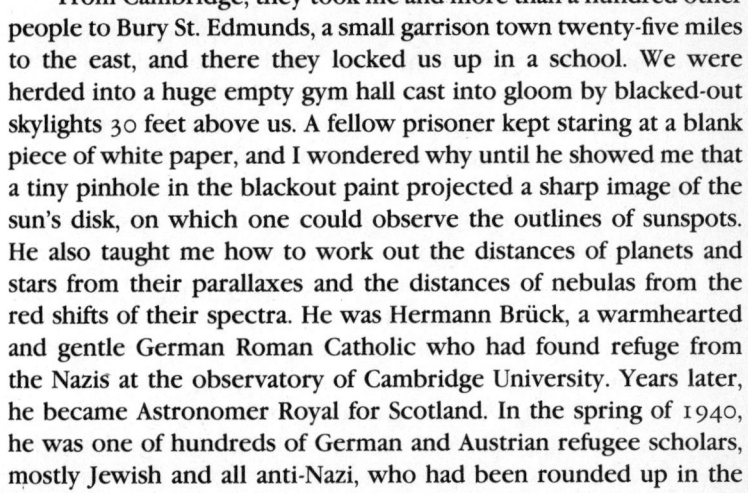

ENEMY
ALIEN

It was a cloudless Sunday morning in May of 1940. The policeman who came to arrest me said that I would be gone for only a few days, but I packed for a long journey. I said good-bye to my parents.

From Cambridge, they took me and more than a hundred other people to Bury St. Edmunds, a small garrison town twenty-five miles to the east, and there they locked us up in a school. We were herded into a huge empty gym hall cast into gloom by blacked-out skylights 30 feet above us. A fellow prisoner kept staring at a blank piece of white paper, and I wondered why until he showed me that a tiny pinhole in the blackout paint projected a sharp image of the sun's disk, on which one could observe the outlines of sunspots. He also taught me how to work out the distances of planets and stars from their parallaxes and the distances of nebulas from the red shifts of their spectra. He was Hermann Brück, a warmhearted and gentle German Roman Catholic who had found refuge from the Nazis at the observatory of Cambridge University. Years later, he became Astronomer Royal for Scotland. In the spring of 1940, he was one of hundreds of German and Austrian refugee scholars, mostly Jewish and all anti-Nazi, who had been rounded up in the

101

official panic created by the German attack on the Low Countries and the imminent threat of an invasion of Britain.

After a week or so at Bury, we were taken to Liverpool and then to an as yet unoccupied housing estate at nearby Huyton, where we camped for some weeks in bleak, empty, semidetached two-story houses, several of us crowded into each bare room, with nothing to do except lament successive Allied defeats and worry whether England could hold out. Our camp commander was a white-mustached veteran of the last war; then, a German had been a German, but now the subtle new distinctions between friend and foe bewildered him. Watching a group of internees with skullcaps and curly side-whiskers arrive at his camp, he mused, "I had no idea there were so many Jews among the Nazis." He pronounced it "Nasis."

Lest we escape to help our mortal enemies, the army next took us to Douglas, a seaside resort on the Isle of Man, where we were quartered in Victorian boardinghouses. I shared my room with two bright German medical researchers, who opened my eyes to the hidden world of living cells—a welcome diversion, lifting my thoughts from my empty stomach. On some days the soldiers took us out for country walks, and we ambled along hedge-flanked lanes two abreast, like girls from a boarding school. One day near the end of June, one of our guards said casually, "The bastards have signed." His terse message signified France's surrender, which left Britain to fight the Germans alone.

A few days later, tight-lipped army doctors came to vaccinate all men under thirty, with the same needle—an ominous event, whose sinister purpose we soon learned. On July 3 we were taken back to Liverpool, and from there we embarked on the large troop-ship *Ettrick* for an unknown destination. About twelve hundred of us were herded together, tier upon tier, in one of its airless holds. Locked up in another hold were German prisoners of war, whom we envied for their army rations. On our second day out, we learned that a German U-boat had sunk another troopship, the *Arandora Star*, which had been crammed with interned Austrian and German refugees and with Italians who were being deported overseas. More than six hundred of the fifteen hundred people aboard were drowned. After that, we were issued life belts.

Suspended like bats from the mess decks' ceilings, row upon row of men swayed to and fro in their hammocks. In heavy seas, their eruptions turned the floors into quagmires emitting a sickening stench. Cockroaches asserted their prior tenancy of the ship. To this revolting scene, Prince Frederick of Prussia, then living in England, restored hygiene and order by recruiting a gang of fellow students with mops and buckets—a public-spirited action that earned him everyone's respect, so that he, grandson of the kaiser and cousin of King George VI, became king of the Jews. Looking every inch a prince, he used his royal standing to persuade the officers in charge that we were not the fifth columnists their War Office instructions made us out to be. The commanding colonel called us scum of the earth all the same, and once, in a temper, ordered his soldiers to set their bayonets upon us. They judged differently and ignored him. One day I passed out with a fever. When I came to, in a clean sick bay that had been established by young German doctors, we were steaming up the broad estuary of the St. Lawrence River, and on July 13 we finally anchored off gleaming white Quebec City. The Canadian army took us to a camp of wooden huts on the citadel high above the town, close to the battlefield where the English General James Wolfe had beaten the French in 1759. The soldiers made us strip naked so they could search us for lice, and they also confiscated all our money and other useful possessions, but I forestalled them by dropping the contents of my wallet out the window of the hut while we were waiting to be searched, and I went around to pick them up the next day, when the soldiers had gone. Sometimes jewels are safest on a scrap heap.

In Canada our status changed from that of internees to that of civilian prisoners of war, entitling us to clothing—navy jackets with a circular red patch on the back—and army rations, which were welcome after our first two days, when we were without food. Even so, the fleshpots of Canada were no consolation for our new status, which made us fear that we would remain interned for the duration of the war and, worse still, that in the event of England's defeat we would be sent back to Germany to be liquidated by Hitler. To have been arrested, interned, and deported as an enemy alien by the English, whom I had regarded as my friends, made me more bitter than to have lost freedom itself. Having first been rejected as a Jew by my native Austria, which I loved, I now found myself rejected as a German by my adopted country. Since we were kept incommunicado at first, I could not know that most of my English friends and scientific colleagues were campaigning to get the anti-Nazi refugees, and especially the many scholars among them, released. I had come to Cambridge from Vienna as a graduate student in

1936 and had begun my life's research work on the structure of proteins. In March of 1940, a few weeks before my arrest, I had proudly won my Ph.D. with a thesis on the crystal structure of hemoglobin—the protein of red blood cells. My parents had joined me in Cambridge shortly before the outbreak of war; I wondered when I would see them again. But most of all I and the more enterprising among my comrades felt frustrated at having to idle away our time instead of helping in the war against Hitler. I never imagined that before long I would be returning to Canada as a free man, engaged in one of the most imaginative and absurd projects of the Second World War.

Our camp offered a majestic panorama of the St. Lawrence and of the lush green country stretching away to the south of it. As one stifling hot, languid day followed another, freedom beckoned from the mountains on the horizon, beyond the U.S. border. I remembered the bishop's advice to Richard II: "My lord, wise men ne'er sit and wail their woes, but presently prevent the ways to wail." How could I escape through the barbed-wire fence? Suppose I surmounted that hurdle without being spotted by the guards, who stood on watchtowers with their machine guns trained on us? Who would hide me after my absence had been discovered at the daily roll call? How could I persuade the Americans to let me join my brother and sister in their country, and not lock me up on Ellis Island? These questions turned over and over in my mind as I lay on my back in the grass at night, listening to the faint hooting of distant trains and watching the delicately colored flashes of the northern lights dance across the sky. Soon I began to dream of jumping on freight wagons in the dark or of fighting my way across the frontier through dense mountain forests—or just of girls.

As a Cambridge Ph.D. of four months' standing, I found myself the doyen of the camp's scholars, and I organized a camp university. Several of my Quebec teaching staff have since risen to fame, though in different ways. The Viennese mathematics student Hermann Bondi, now Sir Hermann, taught a brilliant course in vector analysis. His towering forehead topped by battlements of curly black hair, he arrived at his lectures without any notes and yet solved all his complex examples on the blackboard. Bondi owes his knighthood to his office as chief scientist in Britain's Ministry of Defence, and his fame to the steady state theory of the universe. This theory postulates that, as the universe expands, matter is continuously being created, so that its density in the universe remains constant with time. In this case the universe need not have started with a big bang, because it would never have known a beginning and it would have no end. Bondi developed that ingenious theory with another Viennese interned with us—Thomas Gold, who, like him, was still an undergraduate at Cambridge and who was until recently professor of astronomy at Cornell University. The theory's third author was Fred Hoyle, the Cambridge cosmologist and science-fiction writer. The theory was later disproved by Arno A. Penzias and Robert W. Wilson's discovery of the cosmic microwave background radiation that testifies to the origin of the universe with a big bang.

Theoretical physics was taught to us lucidly by Klaus Fuchs, the tall, austere, aloof son of a German Protestant pastor who had been persecuted by Hitler for being a Social Democrat. Fuchs himself had joined the German Communist party shortly before Hitler came to power and fled to England soon afterward to study physics at Bristol University. After his release from internment in Canada, he was recruited to work for the atomic bomb project, first in Birmingham and then at Los Alamos, and when the war was over he was appointed head of the theoretical physics section of the newly established British Atomic Energy Research Establishment, at Harwell. Everywhere Fuchs was highly regarded for his excellent scientific work, and at Harwell he was also noted for his deep concern with security. Then in the summer of 1949, just before the explosion of the first Russian atomic bomb, the Federal Bureau of Investigation found reason to suspect that a British scientist had passed atomic information to the Russians, and the bureau's description in some ways fitted Fuchs. After several interrogations, Fuchs broke down and confessed—in January of 1950—that from the very start of his work he had passed on to the Russians most of what he knew of the Anglo-American project, including the design of the first plutonium bomb. A few days after Fuchs's conviction for espionage, the prime minister, Clement Attlee, assured Parliament that the security services had repeatedly made "the proper enquiries" about Fuchs and had found nothing to make them suspect him of being a fanatical Communist. Neither had I gathered this during my contacts with him in Canada, but when I recently said so to an old colleague, he told me that Fuchs and he had belonged to the same Communist cell while they were students at Bristol. "The proper enquiries" cannot have been all that searching.

Having no inkling of the tortuous mind that later made Fuchs betray the countries and the friends that had given him shelter, I simply benefited from his excellent teaching. In my own lectures, I showed my students how to unravel the arrangement of atoms in crystals, and I spent the rest of my time trying to learn some of the advanced mathematics that I had missed at school and university.

The curfew was at nine-thirty. The windows of our hut were crossed with barbed wire. Its doors were locked, and buckets were put out. Stacked into double bunks, about a hundred of us tried to sleep in one room where the air could be cut with a knife. In the bunk above me was my closest friend from my student days in Vienna. We had roughed it together in the mosquito-ridden swamps of northern Lapland and had almost suffered shipwreck on a small sealer in the stormy Arctic Ocean. These adventures had inured us to the physical hardships of internment, but the exhilarating sense of freedom that they had instilled in us made our captivity even harder to bear. Lacking other forms of exercise, we made a sport of reading our jailers' regulation-ridden minds. One day the prisoners were told that each could send a postcard to his next of kin in England, but two weeks later all the postcards were returned—without explanation. The camp seethed with frustration and angry rumors, but my friend and I guessed that after leaving the postcards lying around for a couple of weeks, the army censor returned them all because not every card carried its sender's full name. It took a month more before my card reached Cambridge, with the laconic message that Prisoner of War Max Perutz was safe and well.

In time, we learned through rumor that our scenic and efficient camp was to be dismantled and we were to be divided between two other camps. Would friend be separated from friend? By age or by the alphabet? It occurred to me that the pious Quebecois might divide us into believers and heretics—that is to say, into Roman Catholics and the rest—and my hunch was soon confirmed. Since my Viennese friend was a Protestant and I was a Roman Catholic, we were destined for different camps. Adversity tightens friendships. Our familiar Viennese idiom, my friend's keen sense of the ridiculous, and shared memories of carefree student days with girls, skiing and mountain climbing, had helped us to escape from the crowd of strangers around us into our own private world. I decided to stay with the Protestants and the Jews, who also included many scientists, and soon found a Protestant who preferred to join the Catholics. Like Ferrando and Guglielmo, the handsome young swains in *Così Fan Tutte*, we swapped identities. The false Max

Perutz was sent with the faithful to the heaven of a well-appointed army camp, while I was dispatched with the heretics and Jews to the purgatory of a locomotive shed near Sherbrooke, Quebec. To start with, it had five cold-water taps and six latrines for 720 men. We went on hunger strike in protest.

Some weeks later, my comedy of errors was unmasked. The stern camp commander was impressed by the purity of my motives but sentenced me to three days in the local police prison all the same. Here was privacy at last—yet not quite. They locked me up in a cage resembling that of a monkey in an old-fashioned zoo. It had no chair, no bed—only some wooden planks to rest on. Unlike the prisoner in Oscar Wilde's "Ballad of Reading Gaol," I did not look

> *With such a wistful eye*
> *Upon that little tent of blue*
> *Which prisoners call the sky,*
> *And at every drifting cloud that went*
> *With sails of silver by*

because I never even saw the sky. But I had smuggled in several books inside my baggy plus fours, so I was not as bored as the poor soldier who had to march up and down on the other side of the iron grille to guard me. My reading was undisturbed and my sleep interrupted only by the occasional drunk; the little mites burrowed into my skin without waking me. Only when they had made themselves at home there during the weeks that followed did the scabies rash keep me awake at night.

Back in the Sherbrooke camp my spirits sagged at the prospect of wasted years; then the camp commander summoned me again— this time to tell me that my release had been ordered by the British Home Office and that I had also been offered a professorship by the New School for Social Research, in New York City. He then asked me if I wanted to return to England or remain in the camp until my release to the United States could be arranged. I replied that I wanted to return to England, and this drew the admiring comment that I would make a fine soldier. I have never heard that said by anyone else, before or since, but what led me to my decision was that my parents, my girlfriend, and my research were in England, and from the safe distance of Sherbrooke the U-boats and the blitz did not frighten me. My American professorship had been arranged by the Rockefeller Foundation as part of a rescue campaign for the scholars whom the foundation had supported before the war broke out, and in principle it would have qualified me for an American immigration visa, but I was sure that as a prisoner of war without a passport I would never get such a visa. The camp commander raised my hopes that I would be sent home soon.

From our perch on the citadel of Quebec, we had been able to watch the ships go by on the St. Lawrence, but in the locomotive shed we could only watch the men line up for the latrines. In Quebec, we had had a room in a hut set aside for quiet study, but here among a milling, chatting crowd of men my assaults on differential equations petered out in confusion. Camp committees, locked in futile arguments over trivial issues, were chaired by budding lawyers fond of hearing themselves talk. In excruciating boredom, I waited impotently from day to day for permission to leave, but weeks passed and my captivity dragged on. There was little news from home except for hints that my father, who was then sixty-three and had been an Anglophile from youth, had been interned on the Isle of Man. He shared that fate, I learned afterward, with a frail, meticulous old Viennese with sensitively cut features who was distraught at having his life's work interrupted for a second time. This was Otto Deutsch, the author of the then incomplete catalog of Franz Schubert's collected works. He finished it in later years at Cambridge.

Early in December, I was among some prisoners destined for release from my camp and from several other camps who were at last put on a train going east. From its windows, the snow-clad forest looked the same each day, so that we seemed to move merely to stay in the same place, like Alice running with the Red Queen. I had been sad at leaving my Viennese friend behind but was overjoyed to find his father—whom he had feared drowned on the *Arandora Star*—among the prisoners on the train. Some weeks earlier, the father, on discovering that his son was interned in another Canadian camp, had asked to be transferred there, and he was disconsolate that instead the army had now put him on a train carrying him even farther away. The train finally dumped all of us in yet another camp—this one in a forest near Fredericton, New Brunswick. No one told us why or for how long. In the arctic weather, I contracted a bronchial cold that made the dark winter hours seem endless. My father had taught me to regard Jews as champions of tolerant liberalism, but here I was shocked to run into Jews with an outlook as warped and brutal as that of Nazi storm troopers. They were members of the Stern Gang, which later became notorious in Israel for many senseless murders, including that of the Swedish Count Folke Bernadotte, whom the United Nations had appointed as mediator in the Arab-Israeli conflict.

At Christmas, we were finally taken to Halifax, where we were met by one of Britain's prison commissioners—the shrewd and humane Alexander Paterson—sent out by the Home Office to interview any of the internees who wanted to return to Britain. His mission was stimulated by public criticism—"Why Not Lock Up General de Gaulle?" was one of the sarcastic headlines in a London paper that helped to make the War Cabinet change its policy. Paterson explained that it had been impossible to ship any of us home earlier, because the Canadians had insisted that prisoners of war must not be moved without a military escort yet had refused either to release us in Canada or to escort us to England on the ground that our internment was Britain's affair. The British War Office had now fulfilled the letter of the regulation by detailing a single army captain to take us home.

Chaperoned by one urbane captain, 280 of us embarked on the small Belgian liner *Thysville*, which had been requisitioned by the British army complete with its crew, including a superb Chinese cook. From this moment we were treated as passengers, not prisoners, but I became fretful once again when days passed and the *Thysville* had not cast off her moorings; no one had told us that we had to wait for the assembling of a big convoy. As we finally steamed out to sea, I counted more than thirty ships, of all kinds and sizes, spread over a huge area. At first Canadian destroyers escorted us, but we soon passed out of their range, and our remaining escort consisted of only one merchant cruiser—a passenger liner with a few guns on deck—and a single submarine, neither of them a match for the powerful German battleships *Scharnhorst* and *Gneisenau*, which, so our radio told us, prowled the Atlantic not far from our route. We steamed at only nine knots—the speed of the slowest cargo boat—and took a far-northerly course, trusting to the arctic night to hide us. Both my Viennese friend and his father were on board.

Early in the voyage, I stood at the railings imagining a torpedo in every breaker. Like the Ancient Mariner,

Alas! (thought I, and my heart beat loud)
How fast she nears and nears!

But time soon blunted my fears, and I began to enjoy the play of wind and waves. I slept in a warm cabin between clean sheets, took a hot bath, brimful, each morning, ate my meals from white table linen in my friends' company, walked in the bracing air on deck, or retired to read in a quiet saloon. Toward the end of the third week, we were cheered by the sight of large black flying boats of the Coastal Command circling over us, like sheepdogs running round their flock, to keep the U-boats at bay. One gray winter morning, the entire convoy anchored safely in Liverpool harbor. On landing, I was formally released from internment, handed a railway ticket to Cambridge, and told to register with the police there as an enemy alien. When I presented myself at a friend's house near London that night, she found me looking so fit that she thought I must have returned from a holiday cruise, but then she admired the elaborate needlework by which I had kept my tweed jacket in one piece for all those months so as not to have to wear the prisoners' blue jacket with the large red circle on the back. Next morning, at the Cambridge station, our faithful lab mechanic greeted me not as an enemy alien but as a long-lost friend; he brought me the good news that my father had been released from the Isle of Man a few weeks earlier and that both he and my mother were safe in Cambridge. That was in January of 1941.

Less than three years later, I returned to Canada as a representative of the British Admiralty and was accommodated in a suite in Ottawa's luxurious Hotel Château Laurier without being searched for lice. I owed that change of fortune to the remarkable Geoffrey Pyke, former journalist and amateur strategist, who enlisted me for a project that bore the mysterious code name Habakkuk. In 1938 I had taken part in an expedition to the Swiss Alps, where we found out how the tiny snowflakes that fall on a glacier grow into large grains of ice. It had never occurred to me that the expertise I gained there would be of any use to the war effort. When I returned from internment, my professor W. L. Bragg encouraged me to resume my peacetime research on the structure of proteins, with the continued support of the Rockefeller Foundation, and for a long time nobody wanted my help for anything related to the war except fire watching on the roof of the laboratory at night.

At last, one day in the spring of 1942, an urgent telephone call summoned me to London. I was directed to an apartment in Albany—a building owned by the eccentric William Stone, who was also known as the Squire of Piccadilly—where wealthy members of Parliament and writers like Graham Greene rented pieds-à-terre. There I was met by Pyke, a gaunt figure with a long, sallow face, sunken cheeks, fiery eyes, and a graying goatee, who was camped out amid piles of books, journals, and papers, and cigarette butts lying scattered on oddments of furniture. He looked like a secret agent in a spy film and welcomed me with an air of mystery and importance, telling me in a gentle, persuasive voice that he was acting on behalf of Lord Louis Mountbatten, then chief of Combined Operations, to ask my advice about tunneling in glaciers.

Six months went by before Pyke called me again. This time he sized me up with a volley of provocative remarks and then told me, with the air of one great man confiding in another, that he needed my help for the most important project of the war—a project that only he, Mountbatten, and our common friend John Desmond Bernal knew about. When I asked him what it was, he assured me that he would willingly disclose it to me, a friend who had understood and appreciated his ideas from the first, but that he had promised to keep it to himself, lest the enemy or, worse, that collection of fools on whom Churchill had to rely for the conduct of the war should get to hear about it.

I left excited and not much the wiser about what I was supposed to do, but Bernal, who had been my first research supervisor at Cambridge, told me a few days later that I should find ways of making ice stronger and freezing it faster—never mind what for. The project had the highest priority, and I could requisition any help and facilities I needed. Despite my glacier research, I was not sure exactly what the strength of ice was and could find little about it in the literature. Tests soon showed that ice is at the same time brittle and soft, and I found no way of making it stronger. Then one day Pyke handed me a report that he said he found hard to understand. It was by Herman Mark, my former professor of physical chemistry in Vienna, who had lost his post there when the Nazis overran Austria and had found a haven at the Polytechnic Institute of Brooklyn. As an expert on plastics, he knew that many of them are brittle when pure but can be toughened by embedding fibers such as cellulose in them, just as concrete can be reinforced with steel wires. Mark and his assistant, Walter P.

Hohenstein, stirred a little cotton wool or wood pulp—the raw material of newsprint—into water before they froze it and found that these additions strengthened the ice dramatically. When I had read their report, I advised my superiors to scrap our experiments with pure ice and set up a laboratory for the manufacture and testing of reinforced ice.

Combined Operations requisitioned a large meat-storage facility five floors underground beneath Smithfield Market, which lies within sight of St. Paul's Cathedral, and ordered some electrically heated suits, of the type issued to airmen, to keep us warm at 0°F. They detailed some young commandos to work as my technicians, and I invited Kenneth Pascoe, who was then a physics student and is now a lecturer in engineering at Cambridge, to come and help me. We built a big wind tunnel—to freeze the mush of wet wood pulp—and sawed the reinforced ice into blocks. Our tests soon confirmed Mark and Hohenstein's results. Blocks of ice containing as little as 4 percent wood pulp were weight for weight as strong as concrete; in honor of the originator of the project, we called this reinforced ice "pykrete." When we fired a rifle bullet into an upright block of pure ice 2 feet square and 1 foot thick, the block shattered; in pykrete the bullet made a little crater and was embedded without doing any damage. My stock rose, but no one would tell me what pykrete was needed for, except that it was for Habakkuk. The Book of Habakkuk says, "Behold ye among the heathen, and regard, and wonder marvelously: for I will work a work in your days, which ye will not believe, though it be told you," but this failed to solve my riddle.

At one stage, Mountbatten sent Pyke to Canada in aid of Habakkuk with a personal introduction from Winston Churchill to Mackenzie King, the Canadian prime minister. King received Pyke with outstretched arms, saying, "Mr. Chamberlain has sent me such a nice letter about you." He was just one prime minister behind. While Pyke enlisted the Canadians' help, Mountbatten decided to demonstrate the wonders of pykrete to the British Joint Chiefs of Staff. For this show, Pascoe and I prepared small rods of ice and of pykrete that were exactly the same size. We could break the ice rods easily in our hands, but the rods of pykrete stayed in one piece however hard we tried to break them. We also prepared large blocks of each material as targets to be shot at.

So secret was Habakkuk that no one was supposed to know even who I was, lest my nationality (Austria = mountains =

glaciers = ice) or my research record betray it. Pascoe and I worked in the underground meat store while on the upper floors burly Smithfield porters in greasy overalls carried huge carcasses of meat to and from the elevator. They never gave us any of it to supplement our meager ration. One day Pascoe and I were drinking tea at the bottom of the elevator shaft when we heard one of the porters above us saying contemptuously, "A bloody Austrian," and the other porters chiming in: "A bloody Austrian, a bloody Austrian. . . . " There must have been a security leak.

Who was to demonstrate pykrete to the chiefs of staff? Surely not a civilian and a bloody Austrian—an enemy alien at that! It was decided to assign this task to Lieutenant Commander Douglas Grant, who had been an architect in peacetime and who administered Habakkuk. He had not handled pykrete before, but he did have a uniform. I gave him our rods of ice and of pykrete, packed with dry ice into thermos flasks, and also large blocks of ice and of pykrete, and I wished him luck. I waited for news the next day, but none came.

Rationing had hit the small restaurants and tea shops in the city. Pascoe and I therefore used to take the bus down bomb-scarred Fleet Street to the palatial Combined Operations Headquarters in Richmond Terrace, just off Whitehall, where we could get a square meal at an affordable price and could listen to the latest gossip. But that day the entertaining Pyke was still away in Canada, and everybody else seemed to avoid us. After lunch I searched for the normally unruffled Grant and found him in a black mood. When he had handed our little rods of ice and of pykrete around, the old gentlemen had been unable to break either. Next, he had fired a revolver bullet into the block of ice, which duly shattered, but when he fired at the block of pykrete, the bullet rebounded and hit the chief of the Imperial General Staff in the shoulder. The chief was unhurt, but Habakkuk was under a cloud. Worse was to come.

In Pyke's absence, an admiralty committee headed by the chief of naval construction had examined Habakkuk and sent an unenthusiastic memorandum about it to Mountbatten. When this news reached Pyke in Canada, it merely confirmed his disdain for the conservatism of the British establishment, which he epitomized in a derisive motto: "Nothing must ever be done for the first time." He sent back a cable headed HUSH MOST SECRET. CIRCULATION RESTRICTED TO CHIEF OF COMBINED OPERATIONS ONLY! The message read, CHIEF OF NAVAL CONSTRUCTION IS AN OLD WOMAN. SIGNED PYKE. The classification "Hush Most Secret" was normally reserved for operational matters and was therefore treated with respect, but the contents of Pyke's cable quickly reached the ears of its victim. He was an admiral; outraged at having his courage questioned by a mad civilian, he marched into Mountbatten's office and demanded Pyke's instant dismissal. Habakkuk seemed doomed. Then Pyke returned from Canada, elated by the success of his mission and especially by the splendid performance of a prototype that the Canadians had succeeded in launching on Patricia Lake in Alberta. Prototype of what?

Geoffrey Pyke was born in 1893, the son of a Jewish lawyer who died when the boy was five years old and left his family with no money. His mother seems to have quarreled with all her relatives and to have made life hell for her children. She sent her son to Wellington, a snobbish private school attended mainly by sons of army officers, yet she insisted on his wearing the dress and observing the habits of an Orthodox Jew. This made him the victim of persecution and bred in him a contemptuous hatred of the establishment. Although he never finished his schooling, it was possible for him in those days to start studying law at Cambridge. When the First World War broke out, Pyke decided to stop his studies and become a war correspondent. Characteristically, he began his career by persuading the editor of the *Daily Chronicle* to send him to the enemy's capital, Berlin. He bought an American passport from a sailor in the London docks and made his way to Berlin via Denmark, but he was soon caught and was told that he would be shot as a spy. After some time in jail, he was put into an internment camp at Ruhleben instead. Less than a year later, the *Daily Chronicle* appeared with the banner headline "CORRESPONDENT ESCAPES FROM RUHLEBEN." By ingenious and meticulous planning, Pyke and another Englishman, Edward Falk, had made their way to Holland and then back to England.

Confident now that he could solve any problem by hard thinking, Pyke devised an infallible system for making money on the commodities market. At first he succeeded, and in 1924 he used the money to finance a startling new experiment in education. He founded the Malting House School at Cambridge, where children between the ages of two and five were to receive no formal teaching but instead were to be guided to discover knowledge for themselves in purposeful play—"discovery of the idea of discovery." For a time the school flourished, and it became a laboratory where the great child psychologist Susan Isaacs studied the intellectual growth and social development of young children. Pyke's lawyer urged him to endow the school with the fortune he had made on the Metal Exchange, but he had more grandiose educational plans. To finance them, he bought metals on credit through several brokers, keeping each of them in the dark about the full extent of his operations. At one point he cornered as much as a third of the world's supply of tin. Then the day came when Pyke's infallible graphs misled him: prices fell when they should have risen, and Pyke went bankrupt. His school had to close, his marriage broke up, and his health collapsed. He tried journalism again, but no one would print his long articles, and he lived on the charity of friends. In the mid-1930s, recovering, he organized a campaign for sending supplies to the Loyalists in the Spanish Civil War. Later he raised a band of young English volunteers to conduct a clandestine public-opinion poll in Nazi Germany. Its results were to prove to Hitler that the Germans did not want to go to war, but Hitler forestalled an evaluation of Pyke's poll by the invasion of Poland.

Despite his failures Pyke remained unshaken in his faith that he knew how to perform any job better than those whose profession that job happened to be, and as soon as the Second World War broke out he became intent on telling the soldiers how to win it. Initially no one would listen to him, but persistent campaigning and connections in high places brought him an introduction to Mountbatten. In March of 1942, Pyke proposed to the chief of Combined Operations that Allied commando troops be parachuted into the Norwegian mountains to establish a base on the Jostedals-breen, the great glacier plateau, for guerrilla warfare against the German army of occupation—a base from which the commandos would be able to attack nearby towns, factories, hydroelectric stations, and railways. These troops should be equipped with a snow vehicle of Pyke's design, which would allow them to move at lightning speed across glaciers, up and down mountainsides, and through forests. Pyke persuaded Mountbatten that such a force would be invulnerable in its glacier strongholds and would tie down a large German army trying vainly to dislodge it. Despite Churchill's enthusiastic comment "Never in the history of human conflict will so few immobilize so many," the plan was dropped, perhaps because someone had found out that there are no towns, factories, hydroelectric stations, or railways near the Jostedalsbreen. The snow vehicle that Pyke had demanded for the project was meanwhile built by Studebaker and named the Weasel. It proved its worth during the war in France and Russia and afterward conveyed research expeditions safely to the South Pole.

While Pyke was in the United States organizing the manufacture of Weasels, he composed his great thesis on Habakkuk. From New York he sent it in the diplomatic bag to Combined Operations Headquarters in London with a label forbidding anyone other than Mountbatten to open the parcel. Inserted opposite the first page was a green sheet of paper with a quotation from G. K. Chesterton: "Father Brown laid down his cigar and said carefully: 'It isn't that they can't see the solution. It is that they can't see the problem.' " In his accompanying letter Pyke wrote, "The cover name for this ——— project, because of its very nature, and partly because of you, is Habakkuk, *'parce qu'il était capable de tout.'* "

I cannot remember anyone's ever revealing to me officially what Habakkuk stood for, but gradually the secret leaked out, like acid from a rusty can. Pyke foresaw that for several purposes air cover was needed beyond the range of land-based planes. Conventional carriers, he argued, were too small to launch the heavy bombers and fast fighters that would be needed for the invasion of any distant shores. To extend air cover for Allied shipping over the entire Atlantic, floating islands were needed; such islands would allow planes to be flown from the United States to Britain instead of being shipped. They would also facilitate the invasion of Japan. But what material could such islands be made of, since every ton of steel was needed for ships and tanks and guns, and every ton of aluminum for planes? What material existed that was still abundant? To Pyke, the answer was obvious: ice. Any amount of it could be had in the Arctic; an island of ice melts very slowly and could never be sunk. Ice could be manufactured with only 1 percent of the energy needed to make an equivalent weight of steel. Pyke proposed that an iceberg, either natural or artificial, be leveled to provide a runway and hollowed out to shelter aircraft.

Mountbatten told Churchill of Pyke's proposal. Churchill wrote to his chief of staff, General Hastings Ismay:

> I attach the greatest importance to the prompt examination of these ideas. . . . The advantages of a floating island or islands, even if only used as refuelling depots for aircraft, are so dazzling that they do not at the moment need to be discussed. There would be no difficulty in finding a place to put

such a "stepping-stone" in any of the plans of war now under consideration.

The scheme is only possible if we make Nature do nearly all the work for us and use as our materials sea water and low temperature. The scheme will be destroyed if it involves the movement of very large numbers of men and a heavy tonnage of steel or concrete to the remote recesses of the Arctic night.

Something like the following procedure suggests itself to me. Go to an ice field in the far north which is six or seven feet thick but capable of being approached by icebreakers; cut out the pattern of the ice-ship on the surface; bring the right number of pumping appliances to the different sides of the ice-deck; spray salt water on continuously so as to increase the thickness and smooth the surface. As this process goes on the berg will sink lower in the water. There is no reason why at the intermediate stages a trellis-work of steel cable should not be laid to increase the rate of sinking and give stability. The increasing weight and depth of the berg will help to detach the structure from the surrounding ice-deck. It would seem that at least 100 feet in depth should be secured. The necessary passages for oil fuel storage and motive power can be left at the proper stages. At the same time, somewhere on land the outfits of huts, workshops and so forth will be made. When the berg begins to move southwards so that it is clear of the ice floes, vessels can come alongside and put all the equipment, including ample flak, on board.

Could an ice floe thick enough to stand up to the Atlantic waves be built up fast enough? It was to find the answer to this question that Pyke and Bernal first called me in, but they were not allowed to tell me what the question was. As anyone knows who has tried to make a skating rink in his backyard, a long time is needed even in very cold weather to freeze a thick layer of water, because the thin film of ice that forms at the top delays the transfer of heat from the underlying water to the cold air above. By Churchill's method, it would have taken about a year to build up an ice ship 100 feet thick—and then only if the action of natural forces could somehow be prevented from causing it to disintegrate. Hence this method was deemed unfeasible. Then what about a natural floe? In the 1930s a Russian expedition had discovered that even at the North Pole the pack ice was not more than 10 feet thick. Atlantic waves can be as high as 90 feet, with a distance of more than 1,500 feet from crest to crest. Our tests showed that a slab of ice 10 feet thick and suspended on two knife edges would snap in the middle even if the waves were only 800 feet apart. Besides, bombs and torpedoes would crack it even if they could not sink it, and natural icebergs have too small a surface above water for an airfield and are liable to turn over suddenly.

The project would have been abandoned in 1942 if it had not been for the discovery of pykrete: it is much stronger than ice and no heavier; it can be machined like wood and cast into shapes like copper; immersed in warm water, it forms an insulating shell of soggy wood pulp on its surface, which protects the inside from further melting. However, Pascoe and I found one grave snag: although ice is hard to the blow of an ax, it is soft to the continuous pull of gravity, which makes glaciers flow like rivers—faster in the center than at their sides, and faster at the top than near their beds. If a large ship of ordinary ice were kept at the freezing point of water, it would gradually sag under its own weight, like putty; our tests showed that a ship of pykrete would sag more slowly, but not slowly enough, unless it were cooled to a temperature as low as 4°F. To keep the hull that cold, the ship's surface would have to be protected by an insulating skin, and its hold would have to carry a refrigeration plant feeding cold air into an elaborate system of ducts.

All the same, plans went ahead. Experts drew up requirements, naval designers settled in at their drawing boards, and committees held long meetings. The Admiralty wanted the ship to be strong enough to stand up to the biggest known waves—100 feet high and 2,000 feet from crest to crest—even though such gigantic waves had been reported only once, in the north Pacific, after prolonged storms. It said that the ship must be self-propelled, with enough power to prevent its drifting in the strongest gales, and that its hull must be torpedo-proof, which meant that it had to be at least 40 feet thick. The Fleet Air Arm demanded a deck 50 feet above water, 200 feet wide, and 2,000 feet long, to allow heavy bombers to take off. The strategists required a cruising range of 7,000 miles. The final design gave the bergship (as it came to be referred to) a displacement of 2.2 million tons—twenty-six times that of the *Queen Elizabeth*, the biggest ship then afloat. Turbo-electric steam generators were to supply 33,000 horsepower to drive twenty-six electric motors—each fitted with a ship's screw and housed in its own separate nacelle—on the two sides of the hull. These motors were to propel the ship at 7 knots, the minimum speed needed to prevent its drifting in the wind.

Steering presented the most difficult problem. At first we thought that the ship could be steered simply by varying the relative speed of the motors on either side, like a plane taxiing along the ground, but the navy decided that a rudder was essential to keep the ship on course. The problem of suspending and controlling a rudder the height of a fifteen-story building was never solved. Indeed, even today rudders cause problems in supertankers of only a tenth the bergship's tonnage: in 1978 failure of the rudder control caused the supertanker *Amoco Cadiz* to be blown onto the rocks off the coast of Brittany, spilling its oil on the white beaches.

While plans for the bergship became more elaborate with each committee meeting, Pyke's mind raced ahead to work out how such a ship should be used to win the war. He argued that the bergships would solve the difficult problems of invading hostile coasts, because they would be able to force their way straight into the enemy's harbors. The defending troops would be petrified, literally, by being frozen solid. How? The bergships were to carry enormous tanks full of supercooled water—liquid water cooled below its normal freezing point—which could be sprayed at the enemy to solidify on contact. Afterward more supercooled water would be pumped ashore to build bulwarks of ice, behind which Allied troops could safely assemble and make ready to capture the town. It was Pyke's best piece of science fiction. In reality, the cooling of liquid water below its freezing point is observed only in the tiny droplets that clouds are made of. Pyke could not have found reports in the scientific literature of anyone's making more than a thimbleful of supercooled water, but this fact did not diminish his enthusiasm for its use by the ton.

My own next problem was to find a site for building a bergship. How could we follow Churchill's sensible directive to let Nature do the job? Surveying the world's weather maps, I was unable to find a spot on Earth cold enough to freeze 2 million tons of pykrete in one winter. Nature would have to be aided by refrigeration. Eventually we chose Corner Brook, in Newfoundland, where wood pulp provided by the local mills was to be mixed with water and frozen into blocks in a 200-acre refrigeration plant. The problem of launching our leviathan was to be circumvented by laying down the first pykrete blocks on wooden barges cramped together to form a large floating platform. This would gradually sink as the mass of pykrete was built up. The prototype was to be built in the winter of 1943–44, to be followed by a fleet of bergships constructed on the north Pacific coast the following winter, in time for the invasion of Japan.

One day Mountbatten called me into his office to ask who should represent Habakkuk at a high-level meeting. I suggested Bernal as the only man who possessed the technical knowledge, the intellectual stature, and the persuasiveness to stand up to the war leaders. Bernal was the most brilliant talker I have ever encountered. The son of a wealthy Irish Catholic farmer, he soaked up knowledge like blotting paper from an early age and became mesmerized by science. Once he tried to generate X rays by focusing the light from a paraffin lamp so as to see through his hand— he nearly set the farm on fire and was beaten by his father. He was converted to communism in 1922, when he was a student at Cambridge, and remained a faithful Party member all his life. (He died in 1971.) Bernal is mentioned in Andrew Boyle's book *The Climate of Treason*[1] as one of the founders of the Cambridge Communist cell in the 1930s, but he made no secret of his allegiance and was never suspected of disloyalty to Britain. As a Cambridge undergraduate in the early 1920s, he studied the natural sciences and then took up X-ray crystallography, a physical method used for determining the arrangements of atoms in solids. When I joined him as a graduate student, in 1936, he was at the height of his powers, with a wild mane of fair hair (no beard), sparkling eyes, and lively, expressive features. We called him Sage, because he knew everything from physics to the history of art. He was a bohemian, a flamboyant Don Juan, and a restless genius always searching for something more important to do than the work of the moment.

When war broke out, the authorities asked Bernal to assess the likely damage from aerial bombardment. He requested that his former research assistant be taken on to help him, but to his astonishment the request was refused on security grounds. Bernal ridiculed the decision and demanded to see the reason. When he was reluctantly shown the file, the papers stated that the man could not be trusted because he was associated with the notorious Communist Bernal.

Mountbatten, who liked to have unconventional people around him as counterweights to naval orthodoxy, appreciated Bernal's prodigious knowledge and his original approach to any kind of problem. Mountbatten himself impressed me greatly by his quick and decisive mind. The high-level meeting he was preparing for took place in Quebec in August of 1943 and was headed by Roosevelt and Churchill. Bernal staged a demonstration of pykrete, which so impressed the war leaders that they decided to give Habakkuk the highest priority. Detailed plans for the immediate construction of a prototype were to be drawn up in Washington. The British team was ordered there forthwith—except for Pyke, whose mordant wit had upset the American military to the point where he was forbidden to come.

When the people at the U.S. consulate in London saw my invalid Austrian passport, they said that they were not allowed to issue visas to enemy aliens, however vital for the war effort. Mountbatten's chief of staff tackled this trivial obstacle by phoning the Home Office and telling the people there to make me a British subject within the hour. But, like a parson asked to perform a shotgun wedding without calling the banns, the Home Office insisted on at least the semblance of its customary naturalization ritual. That night a detective called on me at my lodgings in Holland Park. Would I give the names of four British-born householders who could vouch for my loyalty? Normally, the detective said, he would make careful inquiries of each of them, but in my case he wouldn't bother. What near relatives did I have in enemy territory? Normally, he would cross-check my answers, but in my case he wouldn't bother. Had I been convicted of any crime? Yes, of riding a bicycle in Cambridge without lights. Normally, he would check the police records, but in my case he wouldn't bother. After an hour of such banter, I signed his form. Supposing I had gone and betrayed the secrets of pykrete to the Eskimos. Would the prime minister have assured Parliament that "the proper enquiries" had been made at the time of my naturalization to ascertain that I had not been an Eskimo sympathizer from an early age?

The next morning I swore allegiance to the king before a justice of the peace; my wife, an ex-German, merely had to sign a piece of paper at home in Cambridge. It seems that the king did not care about the allegiance of women. The following day I was issued a shiny blue passport that described me as a "British Subject by Certificate of Naturalization Issued 3 September 1943." You cannot become an Englishman, as you can become an American, but at least my wife and I were no longer enemy aliens liable to be interned, and my new passport solved the U.S. visa problem.

The other members of the Habakkuk team had already sailed

to New York. To catch up with them, I was now sent there by air. First, a Sunderland flying boat took me from Bournemouth to Shannon, where the British officers on board donned civilian clothes in deference to Eire's neutrality. From Shannon, Pan Am's Yankee Clipper flying boat ferried us to Newfoundland in fourteen hours, and thence to New York harbor, where we landed thirty-four hours after leaving London—a record time. When the immigration officer read in my passport that my British nationality was of exactly four days' standing, he decided to unmask this foreign agent whom the wily British were trying to foist upon their unsuspecting ally and subjected me to a sharp interrogation. When I had told him most of my life history, except for my involuntary sojourn in Canada, he began questioning me about relatives in the United States. A brother. What is his name? When was he born? What does he do? Where does he live? My heart thumped as I remembered that my brother's house had been searched by the FBI when they found out that he had been in correspondence with a prisoner of war in Canada. Would this be in the immigration officer's file? If it was, he gave no sign. He continued. What other relatives? A sister. Where does she live? Prytania Street, New Orleans. Suddenly, his tense face relaxed into a broad grin. "But that's the street where I was born." And I was admitted. No one whose sister lived on Prytania Street, New Orleans, could be a spy.

On arriving in Washington, where I imagined the British team to be busy sixteen hours a day with the planning of the bergship's construction, I was surprised to find them all welcoming me at Union Station in the middle of a weekday afternoon. They wondered what the weather had been like in London when I left—a question that I diagnosed as an expression of homesickness—and seemed in no hurry to get back to their desks. The next morning, when I reported for duty in a hut outside the Department of the Navy Building on Constitution Avenue, I heard that Habakkuk was under scrutiny by the department's naval engineers and that pending their report there was nothing we could do. Lord Zuckerman, another of Mountbatten's wartime scientific advisers, recently explained to me why no one paid much attention to us in Washington. Shortly after our arrival there, Mountbatten left Combined Operations to become commander in chief of the Allied forces in Southeast Asia. Since he had been Habakkuk's principal advocate, its priority took a deep plunge.

So as not to idle away my time, I asked for permission to visit the Canadian physicists and engineers who had carried out tests on ice and pykrete parallel to ours and had built a model ice ship, complete with insulation and refrigeration, on Patricia Lake. It was on this trip that I reentered Canada as a free man, but I evaded my hosts' conventional question about whether this was my first visit. Back in Washington, I took a room in the suburbs, where I listened to a Republican fellow lodger's denunciations of Roosevelt as a greater menace than Hitler. I read in the Library of Congress or went rock climbing on the banks of the Potomac until the United States Navy finally decided that Habakkuk was a false prophet. One reason was the enormous amount of steel needed for the refrigeration plant that was to freeze the pykrete, but the crucial argument was that the rapidly increasing range of land-based aircraft was making floating islands unnecessary. This was the end of Pyke's ingenious project.

It was hard for a civilian to find a place on a ship back to England, but finally I was allocated a berth in a first-class single cabin on the *Queen Elizabeth*, England's newest and fastest liner. When I stepped into my cabin, I found that I shared it with five others. One, a tall, dignified old gentleman, introduced himself as Mr. Coffin, moderator of the Presbyterian Church in the United States of America, and proudly announced that he was going to London to have tea with the queen. To belie his lugubrious name, he entertained the rest of us by day with a great fund of stories; he also kept us awake at night with his loud snoring. The ship carried fourteen thousand American soldiers, sent to join the great armies that were to liberate France the following summer. Under big signs proclaiming NO GAMBLING, piles of dollar bills slid across mess tables in the lounge every few minutes as the great ship heeled over, steering its zigzag course to evade the U-boats. After six days we steamed up the Firth of Clyde, where a large Allied battle fleet lay assembled in the gloomy winter morning, the sinister gray shapes anchored between the dark, cloud-covered mountainsides, lending drama to a scene that looked like a Turner painting of a Scottish loch.

When I reported the demise of Habakkuk to my superior at the Admiralty the next morning, he was not surprised. Pyke was disappointed, but he was already busy on new schemes. One of them was the construction of a gigantic tube from Burma into China; much easier than building a road over the mountains, he argued. Through this tube Allied men, tanks, and guns were to be propelled to China by compressed air, like the pneumatic post in department stores, to help Chiang Kai-shek defeat the Japanese army. Another of Pyke's plans plotted the destruction of the Romanian oil fields, from which Germany derived most of its fuel. In the dark of night, one squadron of planes was to attack the fields with high-explosive and incendiary bombs, while another squadron was to drop a force of commandos nearby, charged with destroying the fields on the ground. How could they penetrate the defenses? Disguised as Romanian firemen, they should capture a fire station and drive into the oil field with its engines, pretending that they were on their way to extinguish the fires started by the air raid but fanning them instead.

I had come to realize some months earlier that construction and navigation of the bergships might prove as difficult as a journey to the moon then seemed to me, yet Habakkuk was one of several apparently impossible projects conceived during the war; in each case the question was not so much of absolute feasibility as of whether the strategic advantages to be gained by carrying out the project were in proportion to the manpower and materials required. In retrospect, it seems surprising that Mountbatten should have taken any of Pyke's projects seriously, but then Mountbatten was the youngest member of the chiefs of staff and headed an organization set up for unconventional warfare. Faced with that task, he liked to attract to his headquarters men who had not been to Staff College and whose ideas were therefore less likely to be anticipated by the enemy—never mind if they wore no socks. In peacetime most of Pyke's ideas would have been discarded as the science fiction they were, but Mountbatten relied for scientific advice on Bernal, without realizing that Bernal's one great failing was a lack of critical judgment. Pyke had the Cartesian's arrogant conviction that an intelligent human being could reason his way through any problem rather than Francis Bacon's humble maxim that "argumentation cannot suffice for the discovery of new work, since the subtlety of Nature is greater many times than the subtlety of argument." I returned to Cambridge, sad at first that my eagerness to help in the war against Hitler had not found a more effective outlet, but later relieved to have worked on a project that at least never killed anyone—not even the chief of the Imperial General Staff.

Until recently, I did not know how and why, four decades ago, the British government had decided to intern and deport many thousands of innocent German and Austrian refugees and Italians living in Britain, and to start releasing them again a few weeks later, long before the danger of a German invasion had receded. I have now read *Collar the Lot!* Peter and Leni Gillman's history of the internment of aliens in Britain, which is based on a scholarly study of official documents that were released thirty years after the events and on interviews with many of the survivors.[2] The book reveals a disheartening story of official callousness, interdepartmental intrigue, newspaper hysteria, public lies, lies told to Parliament and to the governments of the Dominions, and, as John Maynard Keynes said of David Lloyd George, decisions taken on grounds other than the real merits of the case. The book tells also of human suffering, and of a few upright individuals whose compassion turned the tide.

The story begins in the autumn of 1939, when the Home Office and the War Office were anxious to avoid a repetition of the wholesale internment of nearly thirty thousand mostly harmless Germans in squalid prison camps that had taken place during the First World War. The home secretary, Sir John Anderson, therefore established tribunals that classified Germans and Austrians as refugees from Nazi oppression and ordered the internment only of those thought to be loyal to the Nazi regime.

On April 9, 1940, German forces invaded Norway, supposedly helped by a fifth column of Norwegian Nazis and by German spies posing as refugees. A month later the Germans invaded Holland and Belgium, and Winston Churchill replaced Neville Chamberlain as prime minister. Churchill held his first cabinet meeting on May 11. At the insistence of the chiefs of staff, the reluctant Anderson was asked to abandon his enlightened policy and to intern all male Germans and Austrians living near the coasts that were threatened by invasion. A few days later, Sir Nevile Bland, the British ambassador at The Hague, returned to London with alarming stories of treachery by German civilians in Holland. His photograph shows him to be supercilious and vacant, like a figure out of Evelyn Waugh's farcical novels about the British upper class. He realized that his important hour had come, and at the end of May he solemnly warned the nation in a radio broadcast, "It is not the German and Austrian who is found out who is the danger. It is the one, whether man or woman, who is too clever to be found out." Having pondered this profound truth, the chiefs of staff warned the cabinet that "alien refugees [are] a most dangerous source of subversive activity," recommending that all should be interned. "The most ruthless action

should be taken to eliminate any chances of Fifth-Column activities."
On May 24 Churchill told the cabinet that he was in favor of removing all internees from the United Kingdom. Newfoundland and
St. Helena were two of the inhospitable places to which Churchill
proposed we should be banished. General Jan Smuts managed to
do one better by suggesting the Falkland Islands instead. On June
10, when Italy declared war, Churchill ordered the Home Office
to "collar the lot" of Italians living in Britain.

Among four thousand Italians interned during the succeeding
two weeks, and among those supposedly most dangerous ones later
selected for deportation overseas, were H. Savattoni, the banquet
manager at the Savoy Hotel, who had worked there since 1906;
D. Anzani, the secretary of the anti-Fascist Italian League of the
Rights of Man; Piero Salerni, an engineer urgently needed by the
Ministry of Aircraft Production; Alberto Loria, a Jew who had come
to Britain in 1911; and Uberto Limentani, a Dante scholar working
in the Italian service of the British Broadcasting Corporation. All
except Loria and Limentani were drowned on the *Arandora Star*.
Limentani later gave this description of his escape from drowning:

> *E come quei che con lena affannata,*
> *Uscito fuor del pelago a la riva.*
> *Si volge a l'acqua perigliosa e guata.*
> *Così l'animo mio, che ancor fuggiva,*
> *Si volse a diretro a rimirar lo passo*
> *che non lasciò giammai persona viva.*
> (*Inferno*, I.22–27)

> And as he, who with panting breath
> Had escaped from the ocean to the shore,
> Turns and stares back at the perilous waters
> So my fugitive soul
> Turned back to contemplate the pass
> That no one has ever left alive.

In the event most of the interned Italians were sent to the
Isle of Man. As for me, on 30 June I was separated from the
other internees and dispatched, together with a few dozen
young men—bachelors aged about twenty-five or more—to
Liverpool, where I was deposited in front of a great gray-painted

transatlantic liner called the *Arandora Star*. I remembered having seen the same liner, then painted all in white, at anchor in I Giardini in Venice eight years before, when she had been on a cruise around the world. At that time, I said to myself, "How splendid it would be to go on a cruise in that ship!" Now, faced with just that opportunity, the prospect seemed a lot less appealing.

The ship was armed with two small cannons, veritable popguns, one in the bow and the other in the stern. There was barbed wire all over the place. I was not put in the hold but into a cabin two or three levels below deck. Outside this cabin, in which I had to sleep on the floor together with three other internees, there was an English sentry, armed with a rifle and fixed bayonet, who told me that we were being transported to Canada. During the night the ship weighed anchor. Late in the afternoon on the following day, we were allowed up on deck for half an hour for a breath of fresh air. I then saw that we were between Scotland and Ireland, at the point where both coastlines are visible. Looking round the ship, I noted that the lifeboats were in very poor condition: they had obvious holes in them. They had been neglected and did not inspire confidence. During that night, that is between 1 and 2 July, the ship must have rounded the northernmost headland of the Irish coast and headed out into the Atlantic. At six-thirty the following morning, I was dozing when there was a sudden and inexplicable crash. At once I felt that some disaster had occurred because there was a fearful clattering noise, as if everything that *could* overbalance had come hurtling down. Through the crack under the door, I could see that the electric light had suddenly gone out and therefore guessed that the generators were out of action. I asked myself what could have happened, and it occurred to me that the ship might have collided with an iceberg. In fact, we had been torpedoed by a German U-boat. I learned later that we were the victims of a famous U-boat commander, Captain Prien, on his way back from patrolling the Atlantic. Seeing our ship sailing without an escort, he had been unable to resist the temptation of firing a torpedo, which struck us full on, and he had then continued on his course.

There were about eighteen hundred souls on board— Italian, Austrian, and German internees, and, naturally, some hundreds of soldiers escorting us. My three cabin mates vanished at once. However, I remained for a few seconds, groping in the dark, because I recalled having seen some life belts hanging on the walls. I found one, put it on, and then somehow or other found my way up on deck. I could see that there was some panic, but I don't think that I lost my head because in the event there was simply no time to worry. I was always able to act with a certain coolness, that is to say, reflecting before taking each decision. The first thing I did was to climb up to the highest point that I could find, in order to determine whether the ship really was sinking. This showed me that it was tilting over more and more steeply to one side. I saw a sailor lowering a lifeboat into the sea and told myself that my best course would be to try to get aboard that lifeboat. But when I got down to the place, I realized that boarding the lifeboat would be like jumping from the fourth story, and I couldn't bring myself to do it. Only one person succeeded in jumping, and he fractured his skull (although he survived). So I gave up that scheme and made my way along the side deck to see if there was any other way of getting into the sea. Eventually, I found a piece of rope that I thought might suit my purpose, but not satisfied even with this, I went on searching and at last found a rope ladder.

At this juncture I decided to wait quietly for a while, thinking it opportune to put off getting into the sea until the last moment, because in the north Atlantic it can be extremely cold on a morning of cloud and rain even though it was July. After a bit I began to climb down the rope ladder, but on reaching the lower deck I thought that it might be better to stop and make sure that the ship really was sinking. Almost immediately I became aware that the end was imminent, so I went down into the sea. My immediate concern was to swim far enough away to avoid being sucked under with the ship.

The few boats that had been launched were for the most part filled with German sailors who had been interned after capture in South Africa and who knew how to lower lifeboats. In all, there were only five or six boats because, as I was told afterward, those positioned on the side opposite the direction in which the ship was tilting could not be lowered. In any event, no lifeboats were visible. There was already some wreckage in the sea, and I swam toward some object thinking that it might keep me afloat. There was another Italian hanging on to this bit of wood, and I said to him, "Help me to push this farther away from the ship so that we can save ourselves." I asked him his name: the poor chap was called Avignone, and I later found his name on the list of those who drowned. Many of those who managed to get down into the sea froze to death after a few hours in those icy waters.

In the meantime, the ship was sinking fast. Almost fascinated by the sight, I turned continually to look, yet I was anxious to get as far away as possible for fear of being dragged under. In fact, many of those who were too close—some good swimmers among them—were sucked under and not seen again. This great liner of about 12,000 to 15,000 tons listed more and more to one side, thereby throwing hundreds of people into the sea, mainly elderly people who had not attempted to save themselves. At that moment, the seawater clearly got into the boilers, because there was an explosion. Almost at once, as the stern sank, the bow lifted briefly above the waves, and with a frightening noise the liner slid obliquely into the sea, making the water boil over all round. There was wreckage everywhere, and corpses. More than once I became entangled in some floating debris that had either wire or metal spikes sticking out of it. There were also patches of diesel oil that had caught fire, and I therefore found myself in the midst of flames, although these quite naturally burned out very quickly. I reckon that I remained in the sea for about two hours.

At first I tried to clamber onto a sort of seat from the ship, which might have served me as a raft, but I was frustrated in this attempt because it overturned each time I got on it. After some time—possibly an hour and a half—I caught sight of a lifeboat a good way off, perhaps a mile or so. I was only able to catch glimpses of this lifeboat when I was lifted up by successive waves, but I determined to make toward it by clinging to some wreckage and propelling it with the help, once again, of another victim of the shipwreck—I think he was an Irishman, probably one of the soldiers who had been in charge of us. So for a while we helped each other, until finally he left me and swam directly toward the lifeboat without any support. I was never able to discover whether he made it. As for me, I told myself that that piece of wood was my only support and that it would be folly to leave it. Even now, I did not doubt for one moment that they would come and rescue us, and it was perhaps this conviction that kept my spirits high, in a manner of speaking. The curious regularity of the waves brought to mind some verses of Alessandro Manzoni's "*Cinque Maggio*," and I repeated them to myself:

> *Come sul capo al naufrago*
> *L'onda s'avvolve e pesa.*

> As the shipwrecked's head
> Is enveloped and weighed down by the waves.

I thought how true it was that the waves broke over and submerged the castaway's head, and I reflected on the significance of the lines that followed:

> *L'onda su cui del misero,*
> *Alta pur dianzi e tesa,*
> *Scorrea la vista a scernere*
> *Prode remote invan.*

> The wave from whose crest
> The doomed man gazed anxiously but vainly
> For a glimpse of a distant shore.

And then, I asked myself, how did it go on? Ah well, I should have to reread the text of "*Cinque Maggio*" when I got back home.

I struggled on with my piece of wreckage for a while and realized that I was growing feebler. Clearly I could not carry on in that laborious fashion. I should have to let go and swim for the lifeboat. I remember thinking that this was a brave decision, that piece of wood being my only secure hold on life. Now I made one last and prolonged effort, since I was still some way off, and managed to get nearer to the lifeboat. By now I was almost completely exhausted, and I made my one mistake of the entire adventure by shouting *Aiuto* ("Help" in Italian). I found out later that aboard that overloaded and waterlogged lifeboat, which was carrying some 110 to 120 survivors, there was a British army captain who had declared that there was no more room and that from then on only British soldiers should be rescued and taken aboard. However, this view was rejected, so I was told later, by the second in command of the torpedoed liner (the captain went down with his ship), a certain Mr. Tulip, at the helm of the lifeboat, who said, "No, we're at sea and we must rescue all survivors." He it was who ordered that they should take me aboard.

As a matter of fact, I managed, with help from those already in the lifeboat, to hoist myself up and realized then that my lungs were on the point of collapse and that my body had been tried to the limit. Squeezed in between the mass of survivors in the lifeboat and shaking with cold, I asked for something to cover myself with. In reply I was punched on the head and found myself sitting at the bottom of the boat with three or four people on my back. By a stroke of luck, reaching out my hands, I found a sailor's jacket and somehow managed to get it on. My position was extremely uncomfortable not only because of the crush of people above me but also because the level of seawater was slowly rising in the bottom of the boat. We should certainly have sunk in a few hours, as some of the German detainee survivors considerably observed. After two hours, making a great effort, I managed to move into a more comfortable position, and by raising myself I was able to breathe freely like the rest.

The lifeboat commander tried to stay in a spot not too far from the other four or five boats. These we could see, but there was no sign of rescue. I reckon that it was some six hours or so after the torpedoing that we saw a four-engined Sunderland seaplane, which was carrying out a search without yet having found any survivors. Then, after a minute or two, it saw us, fired a Verey light, and vanished. We knew now that help would arrive, but we still had to wait for two hours before, to our great relief, a motor torpedo boat bore over the horizon toward us—a Canadian boat called the *St. Laurent*. There were seven hundred survivors of the eighteen hundred passengers on the *Arandora Star*, and first we had to solve the problem of getting aboard the rescue ship, which was anything but easy. The warship took up a position in the center of the large area over which the survivors were scattered, and each lifeboat had to make its way toward it. When it came to transferring from the lifeboat to the rescue ship, the problem was that the swell caused the deck of the motor torpedo boat to be at one moment about 33 feet higher than the lifeboat and the next instant the opposite. And so, in order to get aboard the warship, we had to seize the exact moment when both vessels were on the same level. In my turn, I managed somehow to accomplish this, and I remember having to move along the deck as quickly as possible—I was of course barefooted—since at that point the deck was scorching hot, probably because it was just above the engine room.

Seven hundred is a great crowd on a small vessel like a motor torpedo boat, and although the sailors did what they could, we passed a very unpleasant night, packed as we were into the between decks, to which we had been forced to descend. I found myself sitting on a seat with dozens of other survivors in one of the sailors' sleeping quarters. I spent the night sitting like that, very hungry. I recall getting a cup of hot chocolate laced, if I am not mistaken, with rum. Otherwise, I was safe and sound. Before we were torpedoed, I had a cold, which must have disappeared during my involuntary swim because I don't remember having it afterward. It was certainly a disagreeable night, aggravated by the somewhat irrational fear which spread among us that we might be torpedoed again.

The next morning, 3 July, we arrived off the Scottish coast and disembarked at Greenock. Two or three of the ship-wrecked survivors had died during the crossing, and others had to be taken to hospital. Just before disembarkation, we were, in fact, asked whether we needed medical treatment. At first I refused, thinking that I was in my normal healthy condition, but after running by chance into one or two of my cabin mates from the *Arandora Star* I was advised to go with them to hospital. I then noticed that my bare feet were rather swollen on account of the freezing conditions of the preceding twenty-four hours. So I took their advice and went along with them, which was very lucky for me, because those who did not go to hospital were embarked for Australia the very next day, and their ship was torpedoed somewhere on route. The ship did not sink, but it must have been a terrifying experience.

So there we were, on the deserted wharf of the port of Greenock, a wretched band of shipwrecked civilians with nobody to take care of us. I had on the sailor's jacket I'd found in the lifeboat, but I was still barefoot. After a while a sort of Red Cross hostel opened, but they could give us no more than a biscuit each. Little by little, the powers that be must have noticed our existence, because around midday some trucks arrived to take us along the Firth of Clyde to a hospital whose location was not known to us at the time but which we subsequently discovered to be the Mearnskirk Emergency Hospital in the vicinity of Glasgow. Covered as I was from head to foot in the naphtha that had spread on the sea after the *Arandora Star* had sunk, I needed above all a bath. Instead, I had to wash myself as best I could with a sponge. Then we were put to bed and could relax at last after the exhausting events of the previous day and night.

We stayed in that hospital for seven or eight days, well fed and cared for. We were the first patients of a hospital constructed for the very purpose of caring for victims of the war. The nurses were especially attentive and, apart from the facts that we had to stay in bed and that there was always a sentry at the door of the dormitory, I believe that we had no complaints about our treatment. We had lost all our personal belongings, and after a week we were given some clothes, which were frankly rather comical, being either too big or too small for us, and some equally useless shoes, as well as certain essentials like razors. On about 11 or 12 July, we climbed aboard a bus that took us to a new internment camp. We traveled right across Scotland, although we did not know then that the harsh building in which we were imprisoned, with its massive walls and surrounding barbed wire, was the Donaldson School Hospital on the outskirts of Edinburgh.

After a few days I was permitted to write to the BBC, and only then did my colleagues in London learn that my name had been mistakenly included on the list of those who had drowned. The BBC had obtained at the outset an order for my release, and this was immediately put into effect, so that on 31 July I was freed—that is, I was escorted by a soldier in a streetcar as far as Princes Street Station and there put on the train for London. I reached London on the evening of 31 July, and on the following morning I resumed my work at the BBC. There I continued to broadcast for the next five years, until the end of September 1945, that is for the duration of the war.[3]

Limentani later became professor of Italian at Cambridge University, where he gave me this account of his experiences.

In obedience to the chiefs of staff's directive, the War Office ordered that those who had survived the torpedoing of the *Arandora Star* be reembarked a few days later on the *Dunera*, a ship bound for Australia. Among those guarding the internees as they boarded the *Dunera* at Liverpool harbor was a young soldier named Merlin Scott. That night he wrote a letter home. "I thought the Italian survivors were treated abominably—and now they've all been sent to sea again," his letter said. "That was the one thing nearly all were dreading, having lost fathers, brothers, etc. the first time. . . . Masses of their stuff—clothes etc. was simply taken away from them and thrown into piles out in the rain and they were allowed only a handful of things. Needless to say, various people, including policemen! started helping themselves to what had been left behind. They were then hounded up the gangway and pushed along with bayonets, with people jeering at them. . . . Masses of telegrams came for them from relatives, nearly all just saying 'Thank God you are safe,' and they were not allowed to see them." The telegrams "had to go to a Censor's Office. . . . Some of them said they had no mail for six weeks." Shortly after the *Dunera* left harbor, a German submarine fired two torpedoes at it, but the *Dunera* happened to change course, and the torpedoes missed the ship by about a hundred yards.

Merlin Scott's father was an assistant undersecretary at the Foreign Office. His son's letter made the rounds of the office and was shown to Lord Halifax, the Foreign Secretary. He forwarded it to Sir John Anderson, the Home Secretary, together with a memorandum expressing concern about the bad effect that such inhumanity would have on public opinion at home and in the United States. Halifax and Anderson won over Chamberlain, who until then had been the chief executor of Churchill's deportation policy, and on 18 July, only a week after Scott had written his letter, Chamberlain persuaded the Cabinet that "persons who were known to be actively hostile to the present regimes in Germany and Italy, or whom for other sufficient reasons it was undesirable to keep in internment, should be released." The Cabinet also agreed that the "internal management, though not the safeguarding," of the internment camps should be transferred from the War Office to the Home Office. The deportations were stopped.

The Canadian government at first stonewalled Paterson's proposal to release in Canada those refugees who did not want to return to England, and the American State Department refused to admit even those refugees who had held immigration visas before they were interned. Early in 1941 Ruth Draper, the great diseuse, gave one of her heartwarming performances in Ottawa for the Canadian Red Cross. Afterward the Prime Minister asked her what Canada could do in return. She told him, "There is a young innocent boy, whom I have known since he was a baby, being held in one of your internment camps behind barbed wire, without offense, without a trial." The Prime Minister ordered the boy's release, and his decision opened the door for others. When I paid my return visit to Canada, in October of 1943, the last internment camp had just been dismantled. The Gillmans' book shows that even in wartime one person's compassion can sometimes prevail against hardened politicians and the military.

As far as I know, historical research has found no substance in the ugly rumors of spying by Germans who posed as refugees, either in Norway or in Holland; nor was there ever a case of a German or Austrian refugee in Britain who aided the enemy. Merlin Scott, whose letter saved so many Italians in Britain from internment and deportation, was killed by the Italians in Libya during the first British advance, early in 1941.

He was the only child of Sir David Montague Douglas Scott, who was not told how his son met his death until forty-four years later, shortly after his ninety-eighth birthday, when he received this letter from a soldier who had served under Merlin.

While remembering the 40th anniversary of V.E. [Victory in Europe] day on 8th May, I recalled the privilege it was to have served with Sir Douglas Montague Scott [Merlin], who was Platoon Commander of the bren-gun carriers, A Company, 2nd Battalion, Rifle Brigade.

Sir Douglas and I were in a bren gun carrier and were called back from O.P. [outpost] duty to go into action near Hell Fire Pass, Egypt. He left my carrier to go into a signal carrier driven by Rifleman Savage. Sergeant Whiteman, who was Platoon Sergeant, travelled in my carrier. We went into attack in line and came under heavy fire and, on being given a signal from Sir Douglas, we had to withdraw. The carriers withdrew except for his carrier. Sgt. Whiteman and myself in our carrier went forward again to investigate, while still under heavy fire, and found that Sir Douglas's driver had been killed and Sir Douglas severely wounded in the chest. We coupled up his carrier with a tow chain to pull him back from the line of fire. In so doing his carrier went into a gun emplacement and we had to de-couple the tow chains. To do this we had to drive towards the enemy lines, turn round and recouple up to Sir Douglas's carrier to enable us to get back to our own lines.

On enquiring the condition of Sir Douglas I was told that he had died on his way to hospital. The battalion Commander sent for Sgt. Whiteman and myself to thank us for what we had done and said that the action would be mentioned for a military medal, the result being that Sgt. Whiteman was awarded the D.C.M. It is sad to say that Sgt. Whiteman was himself killed a few weeks later.

On the early morning of this action Sir Douglas was chatting to me and said that if it hadn't been for the war, he might never have met people like myself.

The reason why I am writing this is that quite possibly I could be the only surviving person left from this action and for many times I have felt that I wanted to pass this first hand knowledge on to you.

Sir Douglas Montague Scott was a very brave and courageous gentleman and it was a great honour and pleasure to have served under him.

Sir David told me that Merlin had been compassionate even as a boy. When Merlin's letter from Liverpool arrived, Sir David had been at the Foreign Office in charge of American affairs, which put him in a good position to warn the Foreign Secretary, Lord Halifax, of the bad effect that the maltreatment of the Italians would have on public opinion in the United States. When I visited him in September 1985, he was blind and chairbound, but someone who had known him a few years earlier described him as the handsomest man she had ever met. Sir David died in August 1986, a few months before his one hundredth birthday. His wife told me that he never got over Merlin's death.

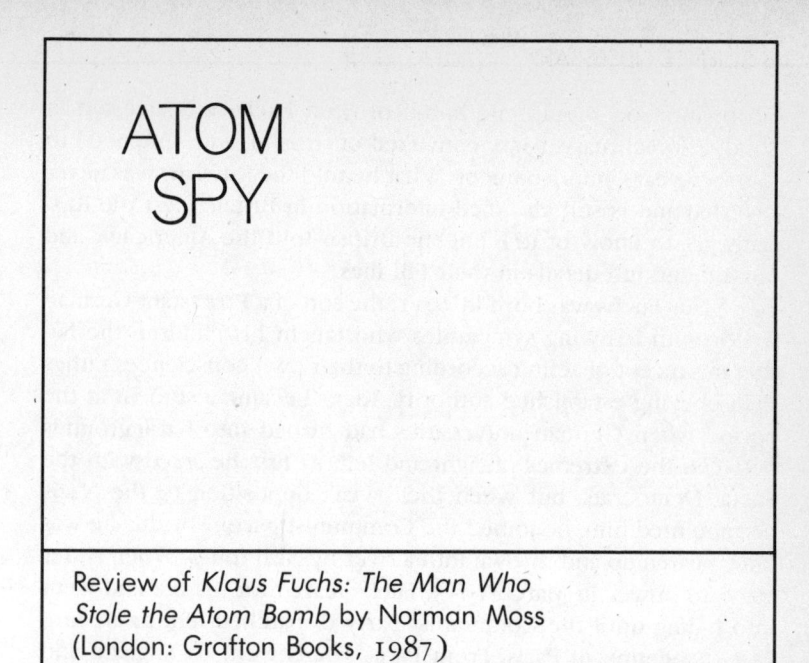

ATOM SPY

Review of *Klaus Fuchs: The Man Who Stole the Atom Bomb* by Norman Moss (London: Grafton Books, 1987)

On 10 September 1949 Michael Perrin, one of the heads of the British Atomic Energy Programme, was woken up by an urgent telephone call asking him to come to the communications room at the U.S. embassy in London. There his opposite number at the Pentagon asked that an RAF plane be sent to the upper atmosphere to check radioactivity detected by the United States Air Force that appeared to signal a Soviet atomic explosion. The public confirmation of this momentous event stunned us. We had believed that Stalin first heard about the American atomic bomb from President Truman at the Potsdam Conference in August 1945, and we could not understand how the Russians had been able to overcome the formidable scientific and technical hurdles involved in the construction of the bomb in no more time than that taken by the cream of European and American physicists, who started in early 1941 and exploded the first bomb in July 1945.

A few weeks before that telephone call, Perrin had received another disturbing piece of news. A coded message sent to Moscow by the Soviet mission in New York during the war and only just deciphered by the U.S. Signal Corps indicated that one Klaus Fuchs, a German-born member of the British team, had given the Russians information on the atomic bomb project. Fuchs was arrested in London in February 1950, convicted of treason, and sentenced to fourteen years' imprisonment. What he told the Russians was never revealed and is still classified information in Britain (lest the Russians get to know of it?), but the British told the Americans, and Moss found full details in their FBI files.

Klaus Fuchs was born in 1911, the son of a Protestant German pastor with left-wing sympathies who taught his children the Lutheran precept of acting according to their own consciences rather than obeying established authority. Klaus became a student at the period when German universities had turned into battlegrounds between the extremes of right and left. At first he sided with the Social Democrats, but when their weak opposition to the Nazis disappointed him, he joined the Communist party. For this he was once beaten up and thrown into a river by Nazi thugs. When Hitler came to power in March 1933, Fuchs feared for his life and went into hiding until the Communist party dispatched him to an anti-Fascist meeting in Paris. From Paris, Fuchs went to England and became a research student in Nevill Mott's great school of theoretical physics at Bristol; after this he worked at Edinburgh with the German physicist Max Born, who was one of the founders of wave mechanics. In the spring of 1940 Fuchs was arrested, interned, and later deported to Canada and released in January 1941, like myself.

The atomic bomb project was set in motion in 1940 by two refugee physicists in Birmingham—the German-born Rudolf Peierls and the Austrian-born Otto Robert Frisch—when they found that the critical mass of the fissile uranium isotope 235 needed for an explosion was no more than a few kilograms. In the summer of 1941 Peierls engaged Fuchs to help him with theoretical work on the project. Nine years later Fuchs confessed: "When I learned the purpose of the work, I decided to inform Russia and established contact through another member of the Communist party." He made this high-handed decision although he had freely signed the Official Secrets Act and thus pledged "that I should not divulge any information gained by me as a result of my appointment to any unauthorised person" and had also applied for naturalization for which, when granted, he would have to "swear by Almighty God that on becoming a British subject I will be faithful and bear true allegiance to His Majesty King George VI, his heirs and successors according to law."

To find out what could have conditioned Fuchs to brush aside these solemn undertakings given freely, I turned to the novelist Arthur Koestler's autobiography, which gives a self-portrait of the author as a young Communist and describes how his normally critical mind came to accept the Party line regardless of any contradictions with reality. Koestler quoted from the German dramatist Bertolt Brecht's play *The Measures Taken*, in which a Greek chorus chants a code of conduct to the young agitators charged with the preparation of a Communist revolution in China.

> Whoever fights for communism
> Must be prepared to fight or not to fight,
> To tell the truth or not to tell the truth,
> To give his services or to refuse them,
> To keep his promises or to break them,
> To be recognized or to be disguised.
> Who fights for communism,
> Has only one single virtue,
> That he fights for communism.

This play was written at the time young Fuchs joined the Party, and even if it did not influence him directly, it probably reflects the prevailing spirit that molded his mentality.

Fuchs was a brilliant mathematician and physicist, also endowed with an accurate memory and a remarkable ability for explaining difficult concepts lucidly. I experienced this clarity when he taught me theoretical physics during our internment at Quebec in the summer of 1940. These talents and the information accessible to him enabled him to provide the Russians with extensive instructions for the manufacture of fissile material and the construction of atomic bombs.

On that first occasion, Fuchs told the Russians that the building of an atomic bomb was possible in principle and that Britain was taking the first steps; he also informed them of the critical mass of a uranium bomb and of his own theoretical work on the separation of the rare fissile isotope of uranium 235 from its more abundant nonfissile partner. Later at Los Alamos, Robert Oppenheimer, the scientific leader of the project, drew Fuchs into scientific discussions on all its aspects, so Fuchs was better informed than most other scientists whose knowledge was restricted to their own special task. Here is an example, in Fuchs's own words, of information handed to a Soviet agent in a town near Los Alamos on one occasion.

Classified data dealing with the whole problem of making an atom bomb from fissionable material as I then knew the problem.

Information as to the principle of the method of detonating an atom bomb.

The possibility of making a plutonium bomb.

The high spontaneous fission rate of plutonium. [It is this that causes fission to occur more quickly than in uranium 235.]

Much of what was then known concerning implosion.

The fact that high explosives as a type of compression was considered but had not been entirely decided upon.

The size as to outer dimensions of the high explosive component.

The principle of the lens system, which had not at that time been finally adopted.

The difficulties of multi-point detonation, as this was the specific problem on which I was then working.

The comparative critical mass of plutonium as compared with uranium 235.

The approximate amount of plutonium necessary for such a bomb.

Some information as to the type of core.

The current ideas as to the need for an initiator.

On another occasion he handed over a report of thirty-five to forty pages and thirteen papers on the separation of the uranium isotopes by diffusion. How easy it all was! General Groves, the American commander of the project, was notorious among the scientists for his obsession with security, but Fuchs could load a pile of secret documents, including a scale drawing of the plutonium bomb, into his car and drive out of the Los Alamos compound to hand them to a Soviet agent in a deserted street of the nearest town. Several such meetings took place in America, and others before and afterward in England.

When Fuchs did not have copies of the relevant documents, he wrote down the mathematical equations and technical data from memory before meeting the agent. He did all this without ever arousing the suspicion of the young English theoretician with whom he shared an office for two years, nor that of any of his other colleagues. Back in England he was appointed head of the theoretical division of the Atomic Energy Research Establishment at Harwell, in which capacity he told the Russians that Britain was building its own atomic bomb even before the cabinet ministers knew of it. As a British civil servant he became a stickler for security. At meetings of the Anglo-American Declassification Committee (also attended by Donald Maclean!), he invariably voted for keeping atomic information classified. As a final irony Fuchs refused to reveal to the Scotland Yard detective who questioned him the technical details he had passed to the Russians, *because the detective was not security cleared*.

When Fuchs was convicted, Parliament and the public wondered how a member of the Communist party could have been allowed to work on the atomic bomb, but Prime Minister Clement Attlee assured the House of Commons:

> Not long after this man came into this country—that was in 1933—it was said that he was a Communist. The source of that information was the Gestapo. At that time the Gestapo accused everybody of being a Communist. When the matter was looked into there was no support for this whatever. And from that time on there was no support. A proper watch was kept at intervals.

When I mentioned this to a veteran physicist friend of mine recently, he interjected, "But Fuchs and I were in the same Communist cell when we were students at Bristol." Born, Fuchs's former chief at Edinburgh, wrote about Fuchs, "He never concealed that he was a convinced communist. During the Russo-Finnish war everyone's sympathies in our department were with the Finns, while Fuchs was passionately pro-Russian." In contrast, Peierls had no idea that Fuchs was a Communist.

When Fuchs finally confessed, he seemed not to realize that he had committed a crime and believed that he could return to Harwell as though nothing had happened. Fuchs had closed his mind to the consequences of his actions, even though he later analyzed it with remarkable insight.

> In the course of this work I began naturally to form bonds of personal friendship and I had concerning them my inner thoughts. I used my Marxist philosophy to establish in my mind two separate compartments. One compartment in which I allowed myself to make friendships, to have personal relations, to help people and to be in all personal ways the kind of man I wanted to be and the kind of man which, in personal ways, I had been before with my friends in or near the Communist Party. I could be free and easy and happy with other people without fear of disclosing myself because I knew that the other compartment would step in if I approached the danger point. I could forget the other compartment and still rely on it. It appeared to me at the time that I had become a "free man" because I had succeeded in the other compartment to establish myself completely independent of the surrounding forces of society. Looking back at it now the best way of expressing it seems to be to call it a controlled schizophrenia.

Fuchs's resolve to decide the fate of the world seems to have been born of an arrogance that bordered on megalomania, as became apparent in prison, where he said to Peierls that when he had "helped the Russians take over everything," he would tell their leaders what was wrong with their system; he did not live up to that promise, because a few years ago he told a Western colleague visiting him in East Germany that Sakharov is a traitor who deserved harsher punishment than mere exile from Moscow.

When Fuchs and I were interned together in Canada, I attended his brilliant lectures, but I had no human contact with that pale, narrow-faced, thin-lipped, austere-looking man; I did not mind that, because I was more interested in the physics he taught me than in the physicist. People who knew him better called him a very cold fish. All Fuchs's former colleagues whom Moss interviewed stressed his reticence. Edward Teller, the Hungarian-born creator of the hydrogen bomb and éminence grise behind President Reagan's Strategic Defense Initiative, found Fuchs "taciturn to a pathological degree." When Teller heard the news of Fuchs's arrest, he is said to have remarked, "So that's what it was!" Another physicist colleague found Fuchs inscrutable, and a woman acquaintance told

Moss that she never heard him laugh. Fuchs had no close friends, never talked about his German past, and appeared to be sexless.

Moss attributes Fuchs's reticence to his desire to conceal his Communist past, but Born's autobiography shows that Fuchs made no secret of it, at least before he started his double life. I suspect that part of his reticence stemmed from his family background: his grandmother, his mother, and one of his sisters all committed suicide in fits of depression; his brother was a consumptive at a hospital in Switzerland; his second sister, who lived in the United States, was hospitalized for schizophrenia in the late 1940s, which suggests that Fuchs must have known her to be mentally unbalanced. Fuchs may have been abnormal in being able to lock his activities into two watertight compartments and to close his mind to the implications of his spying on the colleagues who trusted him and the country that had given him shelter. However, he was no more abnormal than Anthony Blunt, who made friends with the king while spying for Russia.

It has been said that Fuchs handed atomic secrets to an ally and that he merely put into effect the policy of sharing atomic secrets with the Soviet Union that the great Danish physicist Niels Bohr in vain urged upon Roosevelt and Churchill. But there is a world of difference between Bohr's ideal of preventing a nuclear arms race by a policy of mutual trust and Fuchs's secretly giving Russia a good start in the race by handing over to her the scientific results of his colleagues.

Before Fuchs was unmasked, discussions had been under way for continuing the wartime Anglo-American atomic collaboration in peacetime, but afterward, lack of American faith in British security destroyed all prospects of joint work. According to Dean Acheson, who was secretary of state at the time, "The talks with the British and Canadians returned to square one, where there was a deep freeze from which they did not return in my time." Fuchs must have been pleased to have rendered this further gratuitous service to Russia.

How much sooner Fuchs put atomic weapons into Stalin's hands is anyone's guess. Russian physicists were working on the possibility of making an atomic bomb before they received Fuchs's first report; Moss quotes David Holloway's book *The Soviet Union and the Arms Race* for evidence that Fuchs's report made Stalin order the project to proceed with the highest priority.[1] He quotes the estimate of the former head of the Institute of Nuclear Research in East Germany that Fuchs saved the Russians two years; Russian physicists told Peierls that Fuchs saved them between one and two years, but I am skeptical of such estimates. Scientific research is an imaginative activity dependent on qualities of the human mind that are beyond our comprehension. There is no linear progression from the appearance of a problem to its solution. For example, during the Second World War the Germans failed even to start a sustained chain reaction in a nuclear pile because they had determined one of the vital parameters incorrectly. Therefore, it might equally well have taken the Russians many more years to solve the very difficult problems involved if Fuchs had not presented them with the solutions.

What difference that lag would have made politically is even harder to guess. The author quotes evidence that Stalin would not have incited North Korea to attack South Korea if he had not felt secure in the possession of atomic weapons. Would an American monopoly of atomic weapons have deterred the Russians from invading Hungary in 1956? We shall never know.

In the late 1940s Fuchs began to have doubts about Russian policy, which he expressed in his confession.

> It is impossible to give definite incidents [causing his doubts] because now the control mechanism acted against me, also keeping away from me facts which I could not look in the face, but they did penetrate and eventually I came to a point when I knew I disapproved of a great many actions of the Russian Government and of the Communist Party, but I still believed that they would build a new world and that one day I would take part in it and that on that day I would also have to stand up and say to them that there are things which they are doing wrong.

He also began to realize the values that made Britain prevail against the Nazis.

> Before I joined the project most of the English people with whom I had made personal contacts were left-wing, and affected, to some degree or other, by the same kind of philosophy. Since coming to Harwell I have met English people of all kinds, and I have come to see in many of them a deep-rooted firmness which enables them to lead a decent way of

life. I do not know where this springs from and I don't think they do, but it is there.

On being released from Wakefield Prison in 1959, Fuchs went to join his father in East Germany, where he married and was a much honored citizen no longer plagued by doubts about the Communist Party line or nostalgia for the British way of life. The shutters in his mind seemed to have come down once more and stayed. He died in 1987.

GREAT
SCIENTISTS

DISCOVERERS OF PENICILLIN

Review of *Alexander Fleming:
The Man and the Myth* by Gwyn Macfarlane
(London: Chatto & Windus, The Hogarth Press, 1984)

Few people know that in parallel with the race to produce a bomb that would kill people by the hundred thousand, scientists in England ran another race, a race to produce a drug that was to save the lives of millions. This book is the story of the eccentric Scotsman and his chance discovery that set the race in motion. His name was Alexander Fleming, and he was a bacteriologist at a London hospital who in 1929 found that a culture plate seeded with staphylococci had become contaminated with a mold. Instead of discarding it, as others might have done, he noticed something unusual: colonies of cocci had grown everywhere except near the mold, where he saw a clear patch. He cultured the mold and discovered that the broth filtered from it stopped the growth of several kinds of deadly bacteria. Publication of his discovery in a scientific journal stirred up hardly a ripple, and he did little more about it.

Nine years later Ernst Chain, a young German biochemist at the University of Oxford, came across Fleming's paper and decided together with his professor, the Australian pathologist Howard Florey, to find out the nature of the active substance in Fleming's broth and the way it stops the growth of bacteria. With herculean labor, Chain and several of his colleagues extracted a minute quantity of

what they believed to be the pure substance from gallons of broth
and gave it to Florey to test. In May 1940, while the defeated British
army was being evacuated from France, Florey and his team announced a brilliant victory: Florey had injected lethal numbers of
streptococci into eight mice. He had then injected the mold extract
into four of them and left the other four untreated; the treated mice
remained healthy and the untreated ones were dead the next day.
These were the first steps in the purification of penicillin and the
realization of its unparalleled therapeutic powers. As a result Fleming became a world hero, while the names of Florey, Chain, and
their colleagues have remained unknown outside the world of science.

Macfarlane has written a scientific thriller tracing the almost
unbelievable combination of events that guided Fleming to his discovery. He then shows why no one, not even Fleming himself,
recognized the importance of his discovery and why Fleming, even
though he had abandoned the discovery to others, reaped nearly
all the fame. Macfarlane's book is the companion volume to his
earlier, brilliant biography of Howard Florey,[1] and it is equally en-
thralling.

Alec Fleming was the seventh of eight children in a close-knit,
happy family that lived in a farm 800 feet up on a bare, windswept
Scottish hillside. Macfarlane describes the setting of Fleming's childhood with a loving wealth of vivid detail, as though it reflected
his own. Since the eldest Fleming son was to inherit the farm,
the younger brothers sought their fortune in London, because "at
home Scots would have to compete with other ambitious and well-
educated Scots, while in England most of their competitors would
merely be English." When Alec was thirteen, his eldest half brother,
Tom, was already established in London as a prosperous oculist;
Alec followed him, went to a commercial school, and then found
a job as a junior clerk with the America Line. The outbreak of the
Boer War induced the Fleming brothers to volunteer for the London
Scottish Regiment. They were never sent to South Africa, but Alec
became skilled at rifle shooting and once played in a water polo
match against a team from St. Mary's Hospital Medical School.

A friend and colleague of Fleming's said at his funeral:

Looking back on his career, we find woven into the web of his
life a number of apparently irrelevant chance events without
any one of which it would probably not have reached its cli-

max. His choice of a profession, his selection of a medical school, his deviation into bacteriology, his meeting with Alm-roth Wright, the nature of the work he did with him, the chance fall of a tear, the chance fall of a mould, all these events were surely not due to mere chance. We can almost see the finger of God pointed to the direction his career should take at every turn.

That water polo match was the first of these apparently irrelevant chance events, because shortly afterward the prospect of a legacy of £250 made Alec decide to leave his boring job and follow his elder brother into medicine. Which of the twelve London teaching hospitals should he enter? Obviously, after that match, St. Mary's.

All the Fleming brothers seem to have been outstandingly able. In Alec's "selective, penetrating mind all essential facts were ar-ranged in a logical and memorizable order"; he retained effortlessly the mountains of detail and dictionaries full of Greek and Latin terms required for medical examinations, and he carried away all the prizes. At the end of his studies he could have applied for an intern's job at any of the London hospitals, but it so happened that he was one of the stars of St. Mary's Rifle Club. His departure would have spoiled the team's chances of winning a competition for a much-prized cup, and he was therefore recruited as a junior assistant in St. Mary's Inoculation Department. He entered the department just to earn money for his keep, but he was to remain there for the next forty-nine years.

The department was dominated by the formidable Sir Almroth Wright, son of a Yorkshire curate, who was "brought up in an atmosphere of intellectual riches and material poverty entirely fa-vourable to the growth of the life of the mind." When he was a young doctor, his imagination had been fired by Pasteur's vacci-nation against rabies, and he set out to extend it to other diseases. A trial on himself of his first vaccine, against Malta fever, nearly killed him, but his second vaccine, against typhoid fever, heroically tried once more on himself, worked. Despite its immediate success, the Army Medical Corps would not allow Wright to vaccinate more than a few of the soldiers sent to the Boer War, "because all new ideas tend to provoke automatic rejection by an entrenched and conservative profession," with the result that typhoid killed more soldiers than the enemy. Wright was convinced that vaccination could not only prevent but cure many of the bacterial diseases for which there was no effective treatment, and for the rest of his life he worked wrongly and obstinately to prove that point. The Inoculation Department at St. Mary's Hospital was his creation and provided the setting for Fleming's work. It was accommodated in small, overcrowded rooms, but it was a happy place because "shared discomforts endured for the sake of a common purpose tend to promote easy personal relationships rather than friction."

Fleming's specialization in bacteriology was brought about by the First World War, when Wright and his assistants were posted to a grim military hospital in Boulogne. Ever since the days of Lord Lister, the British medical establishment had believed in the efficacy of antiseptics in the treatment of wounds, but Wright and his team found that antiseptics made the soldiers' wounds fester even more, and they "set out to discover why it was that chemicals that would kill bacteria in a test tube in a few minutes, failed to do so in a wound." This question led Fleming to his first important piece of research. He identified the bacteria responsible for wound infections, and he demonstrated elegantly that in an open wound antiseptics like carbolic acid killed the white blood cells that constitute the body's own defense and let the bacteria that had buried themselves in the tissues survive. On the strength of Fleming's findings, Wright campaigned for a change in the treatment of wounds but was again defeated by the conservative medical establishment. I was aghast to read that doctors continued to inject antiseptics into patients with septicemia (blood poisoning) and even covered the noses and mouths of their hapless tuberculosis patients with masks soaked in creosote in the misguided belief that this vile compound would kill the bacteria in their lungs: it was another horrifying example of doctors' determination to inflict tortures on their patients for the sake of the manifestly useless treatments with which the annals of medicine are replete.

The experience of the appalling fate of the wounded in Boulogne generated in Fleming's mind the need for an antiseptic that penetrates the wound and leaves the beneficial white blood cells alive. Tragically, his first treatment of wounds with penicillin suggested that it fulfills only the second condition, otherwise penicillin might have become available twelve years sooner.

When the war ended, Fleming continued his bacteriological work at St. Mary's. Three years later he made the first of his two observations that were to change medical history. Fleming's notebooks, now at the British Museum, fail to describe how it happened; Macfarlane therefore quotes the account of an eyewitness:

> Early on, Fleming began to tease me about my excessive tidiness in the laboratory. At the end of each day's work I cleaned my bench, put it in order for the next day and discarded tubes and culture plates for which I had no further use. He, for his part, kept his cultures . . . for two or three weeks until his bench was overcrowded with 40 or 50 cultures. He would then discard them, first of all looking at them individually to see whether anything interesting or unusual had developed. I took his teasing in the spirit in which it was given. However, the sequel was to prove how right he was, for if he had been as tidy as he thought I was, he would never have made his two great discoveries—lysozyme and penicillin.
>
> Discarding his cultures one evening, he examined one for some time, showed it to me and said "This is interesting." The plate was one on which he had cultured mucus from his nose some two weeks earlier, when suffering from a cold. The plate was covered with golden-yellow colonies of bacteria, obviously harmless contaminants deriving from the air or dust of the laboratory, or blown in through the window from the air in Praed Street. The remarkable feature of this plate was that in the vicinity of the blob of nasal mucus there were no bacteria; further away was another zone in which the bacteria had grown but had become translucent, glassy and lifeless in appearance; beyond this again were the fully grown, typical opaque colonies. Obviously something had diffused from the nasal mucus to prevent the germs from growing near the mucus, and beyond this zone to kill and dissolve bacteria already grown.

Fleming found that saliva, tears, and the whites of eggs also dissolve the bacteria, and that these fluids do no harm to white blood cells, which made him wonder if animals themselves make the ideal antiseptic for which everyone had been searching. This proved a false hope, because Fleming soon found that his nasal mucus or tears left the common disease-producing bacteria unharmed; the bacteria that had dissolved proved to be of a unique kind, blown into his room from no one knew where. Wright, who was a classical scholar, named them *Bacterium lysodeicticus* and the unknown agent in Fleming's nose *lysozyme*. Fleming continued to study this substance for many years, even after his discovery of penicillin, in the hope that it might prove of therapeutic value, but not knowing biochemistry, he never found out what it was or how it worked. This omission was to prove crucial for later developments.

If the discovery of lysozyme was made possible only by the adventitious landing on Fleming's bench of a rare and hitherto unknown germ, the discovery of penicillin resulted from a combination of circumstances improbable beyond belief, of which Fleming's own terse description in the *British Journal of Experimental Pathology* of 1929 provides no hint:

> While working with staphylococcus variants, a number of culture-plates were set aside on the laboratory bench and examined from time to time. In the examinations these plates were necessarily exposed to the air and they became contaminated with various micro-organisms. It was noticed that around a large colony of a contaminating mould the staphylococcus colonies became transparent and were obviously undergoing lysis [being broken up].

Bacteriologists normally grow microorganisms by cooking a nutrient broth, pouring it on a round dish 4 inches wide, letting it solidify into a jelly, stabbing the jelly many times over with a platinum wire dipped into an earlier culture of the microorganism, and finally heating the dish for a day or so in an incubator kept at body temperature. Many years after the event, Ronald Hare, who had been an assistant in the Inoculation Department at the time, tried to rediscover penicillin by preparing a culture plate seeded with staphylococci in just this way and then contaminating it with Fleming's mold.[2] The mold had no effect! To produce the clear patch that Fleming had found, Hare had to seed the dish with the mold *before* seeding it with the staphylococci, but there he encountered another difficulty: the mold would not grow at body temperature! So what could have happened?

In 1928 Fleming was asked to write a chapter on staphylococci for a handbook, and for this purpose he tried to reproduce some anomalous strains reported in the literature; he was helped by a student, D. M. Pryce. Apparently Fleming was in the habit of preparing many bacterial cultures and leaving them scattered on his bench. Pryce told Hare that before going on vacation, Fleming had pushed all those cultures together into a corner to give Pryce space to work in. Pryce himself later went on vacation with Hare, and Fleming returned before they did. When Pryce came back Fleming had piled up his cultures on a tray of antiseptic. Fleming picked up some of them and showed them to Pryce, who remembers that "he took one plate up, looked at it, and after a while said 'That's funny.' " That was the now famous plate referred to in the opening paragraph of Fleming's paper.

To explain what happened, Hare excavated the temperatures recorded in London during the summer of 1928 and assumed that instead of incubating his cultures before going on vacation, Fleming had just left them on his bench. The record shows that the August temperatures were in the sixties, favorable to the growth of only the mold, after which they rose to the seventies, suitable for the growth of the cocci; the cocci grew everywhere except near the mold, which must have exuded a substance inhibiting the multiplication of the cocci. But where did the mold come from? It turned out to be a very rare organism, unlikely to have flown in through the window, which in any case Fleming seldom opened.

Some years earlier a Dutch allergist had given lectures in London advancing the now accepted theory that some patients suffer from asthma because they are allergic to molds. As a result, Wright appointed a young Irish mycologist, C. J. La Touche, to isolate molds from houses inhabited by asthma patients so that the molds could be identified and extracts made from them to desensitize the patients. Because molds produce myriads of airborne spores, mycologists normally grow them under hoods fitted with extract filters, but Wright was a protagonist of the British string and sealing wax tradition of research, and he made La Touche grow his molds in an open makeshift laboratory furnished only with tables. It so happened that this laboratory was immediately underneath Fleming's. It was La Touche who identified Fleming's mold as belonging to the *Penicillium* family; La Touche is not sure that it came from his collection, but both Hare and Macfarlane regard that as the most likely source. Hare concluded,

> Such, then, is what I conceive to be the background to the discovery of penicillin. An accidental observation it is true, but what an accident, depending as it did on a whole series of apparently unrelated events. The choice of Fleming to write a

chapter in a book; the publication of a paper in a scientific journal that prompted him to enquire further; lectures by a Dutch physician that led to the appointment of a mycologist; his working in a laboratory directly beneath that of Fleming; his having the good fortune to isolate a powerful penicillin-producing strain of the mould; his having inadequate apparatus so that the atmosphere became loaded with spores; the high probability that Fleming either forgot to incubate his culture plate or purposely omitted to do so; the fact that Fleming's own laboratory was peculiarly sensitive to outside temperatures; that a cold spell came at a time of the year which is usually unsuitable for the discovery; the visit to Fleming by Pryce that led the former to look again at a plate he had already inspected and discarded; and its having escaped destruction because of entirely inadequate methods for the disposal of used culture plates. All these events, acting in concert, brought to Fleming's notice a phenomenon that cannot, even now, be reproduced unless the conditions in which the experiment is carried out are exactly right. Had only one link in this chain been broken, Fleming would have missed his opportunity.

And if, as Paul Ehrlich used to say, scientific discovery depends partly on *Geld* or money, partly on *Geduld* or patience, partly on *Geschick* or skill and partly on *Glück* or luck, it was the last of them that was almost entirely responsible for the discovery of penicillin. It was, surely, the supreme example in all scientific history, of the part that luck may play in the advancement of knowledge.[3]

Hare omits to mention that luck thrives only in a culture medium containing an ingredient of untidiness, hence the discovery could never have been made in Ehrlich's German laboratory.

Fleming's first announcement to his colleagues at St. Mary's was anticlimactic, as was his subsequent lecture to a wider audience. Again according to Hare,

He had evolved the custom of indulging in a morning potter which included a visit to the big laboratory in which I was working. Fleming's idea of gossip was different from that of most other people. It usually involved his planting himself in front of the fireplace with his hands in his pockets, a cigarette dangling from his lips and looking more or less into space. On rare occasions he would give utterance, usually in the fewest possible words. The information doled out so meagerly might concern anything; that so-and-so had died; that what's-his-name had made a fool of himself again; or how are your Snia Viscosa shares doing? . . .

On this occasion it was neither Fleming's notion of gossip nor high finance that emerged. It was the now famous culture plate that led to the discovery of penicillin. The rest of us, being engaged in researches that seemed far more important than a contaminated culture plate, merely glanced at it, thought that it was no more than another wonder of nature that Fleming seemed to be forever unearthing, and promptly forgot all about it.

Fleming found that his broth inhibited the growth of the streptococci and staphylococci that infected wounds, and also of the organisms responsible for gonorrhea, meningitis, and diphtheria, but not the growth of typhoid, paratyphoid, anthrax, and *Hemophilus influenzae*—also called Pfeiffer's bacillus, the organism that was (incorrectly) believed to have caused the great influenza epidemic after the First World War. The broth was harmless to white blood cells; it could be injected with impunity into mice and rabbits and the mold itself eaten without ill effects.

Having done these experiments, Fleming failed unaccountably to take the obvious next step, the step that Florey was to take twelve years later, namely to find out whether an injection of his broth would protect mice from lethal infection. (Macfarlane calculates that the concentration of penicillin in Fleming's broth would actually have been sufficient to make the experiment work.) Instead, Fleming applied the broth externally to a few patients, with mixed results, and got discouraged. Even more disheartening were the results of two young doctors who tried to extract the active principle from the broth; they found that it lost its activity quickly if kept in water or alcohol, and even more rapidly in blood (a finding that was later proved incorrect); on injection into animals it was eliminated in the urine in less than two hours, while its antibacterial activity took about four hours to develop. To Florey's team, it was like filling a bath with the plug out. Fleming gained the disappointing impression that penicillin would be of no more clinical value than lysozyme, but he found another use for it, and that use is the only one mentioned in his first lecture on penicillin, which bore the unbelievably low-key title "A Medium for the Isolation of Pfeiffer's Bacillus."

The Inoculation Department subsisted not on the charities that in those days supported St. Mary's Hospital but, like the Pasteur Institute in Paris, on the sale of vaccines. Their manufacture had been pioneered by Wright, who had now passed this responsibility on to Fleming. When preparing vaccines against Pfeiffer's bacillus, it was difficult to keep the cultures clear of the ubiquitous staphylococci; Fleming found that addition of penicillin did so. This minor technical advance formed the subject of Fleming's lecture to the Medical Research Club in London, attended by Sir Henry Dale, a leading physiologist and astute director of the National Institute for Medical Research, which had been founded for the specific purpose of developing chemotherapy. Fleming's talk fell flat and aroused no comments or discussion, because, according to Hare, "he was one of the worst lecturers I ever heard, unable to express himself clearly. . . . In a dull monotone, without humour or emphasis, they may have given the audience the facts." Macfarlane recalls that Fleming was often inaudible and, "worse still, gave the impression that he had little enthusiasm for his own subject."[4]

However, Fleming's use of penicillin had one vital consequence: he ordered that henceforth the broth be produced in the Inoculation Department in weekly batches, and he gave subcultures of his *Penicillium* to several colleagues in other laboratories, including Georges Dreyer, the professor of pathology at Oxford whose collaborator, Miss Campbell-Renton, kept the cultures going for the next ten years. Fleming himself made no more mention of penicillin in any of the twenty-seven papers and lectures he published between 1930 and 1940, even when his subject was germicides. Penicillin might thus have been forgotten but for Fleming's earlier discovery of lysozyme.

Howard Florey was an Australian who had migrated to England in 1922, and on Dreyer's death he was appointed professor of pathology at Oxford. Macfarlane, who knew both Fleming and Florey, wrote that there could not have been a greater contrast between the two men. Both were extremely able, but while Fleming was easygoing and laconic, had little ambition, and was popular, Florey was "taut like a coiled spring," worked like a dynamo, and made enemies. Florey suffered from chronic indigestion, which aroused his interest in the composition of mucus secreted by the gut and other tissues. After his arrival at Oxford, he engaged a young German refugee biochemist, Ernst Chain, and, since lysozyme is found in mucus, Florey suggested to Chain that he find out the biochemical mechanism of lysozyme's attack on bacteria. Chain soon solved this problem and then wondered whether lysozyme might not be just one representative of a large class of bactericidal substances occurring in Nature. He collected references to about two hundred papers, going back as far as Pasteur, who was the first to point to the great therapeutic possibilities of "bacterial antagonism."

Many years afterward Chain wrote, "When I saw Fleming's paper for the first time, I thought he had discovered a sort of mould lysozyme."[5] He can't have read that paper very carefully, because it states that penicillin dissolves in alcohol, whereas lysozyme is a protein, and all proteins are insoluble in alcohol. Perhaps this was another one of those strokes of luck without which penicillin would have remained in obscurity, because Chain then proposed and Florey agreed to make penicillin part of a thorough study of natural antibacterial substances.[6] Not that Florey should have needed Chain's prompting. Florey had been a member of the editorial board of the *British Journal of Experimental Pathology* when Fleming's paper appeared there and should have noticed its importance. What is more, he had come to Oxford from Sheffield, where a pathologist grew a culture of Fleming's mold in Florey's own department and used the broth to cure two babies with gonococcal infections of the eyes, as well as a colliery manager with a pneumococcal infection in an injured eye. According to Macfarlane, the pathologist got no encouragement from Florey and gave up when the molds ceased to produce penicillin.

To Chain and Florey therefore belongs the credit for initiating the work that led to the isolation of penicillin and to its clinical trials, even though Chain admitted later that they had done the work for purely scientific rather than medical reasons. To get the mold, Chain merely had to go down the corridor to Miss Campbell-Renton. The heroic efforts needed to produce even a few thousandths of a gram of penicillin from the mold broth are vividly described in Macfarlane's present book and also in his earlier biography of Florey. The instability of penicillin in solution, which had dogged Fleming's collaborators and several others who followed them, was overcome by the newly invented technique of freeze-drying. Having discovered that penicillin could cure lethally infected mice, Florey was determined to find out what it could do in men, but to do so he needed a three thousand times larger dose, requiring 2,000 liters of mold filtrate. He and his team worked night and day to turn their university laboratory into a factory. When wartime shortages cut off the supply of large shallow dishes to culture the molds in, Florey improvised with hundreds of bedpans. To augment his grant from the British Medical Research Council he appealed to the Rockefeller Foundation in New York for support.

When publication of the spectacular results of his first clinical trials in *The Lancet* failed to persuade the British authorities and pharmaceutical firms to put a major effort into the production of penicillin, he went to the United States, where he found what he wanted: enthusiasm, money, know-how, and professionalism. Florey's mission set in motion a vast machine that soon produced penicillin on a scale thousands of times greater than Florey's own "factory" and in much purer form. Chain's 0.005 gram of broth concentrate injected into each of Florey's four mice was later found to have contained no more than 1 part of penicillin in 300 parts of impurities, so Florey did his test with only 1/1,500,000 gram of penicillin for each mouse. By 1945 American production had risen to 100 kilograms a month of pure crystalline penicillin, enough to treat all Allied war casualties with the ideal antiseptic that Fleming had dreamed of in 1919.

How did Fleming react to the success of the Oxford team? He came to visit them after having read their first paper in *The Lancet*, to the surprise of some of them, it was said, because they did not realize he was still alive. He inspected their factory, took in everything he saw and heard, and characteristically said nothing at all, nor did they hear any more from him until the results of their first clinical trials elicited an editorial in the *British Medical Journal* suggesting that the great therapeutic possibilities of penicillin had previously been unrecognized. This drew a refutation from Fleming, quoting a sentence from the summary of his 1929 paper: "It is suggested that it may be an efficient antiseptic for application to, or injection into, areas infected with penicillin-sensitive microbes." Fair enough!

Soon afterward, fame blew in through Fleming's door like one of his microbes. After the publication of the second clinical trial by Howard Florey and his wife, Ethel, *The Times* of London carried a leader about the Oxford work but without mentioning names. Sir Almroth Wright, by now eighty-one years old, thereupon wrote a letter to *The Times* pointing out that the laurel wreath should be placed on Fleming's brow, because he had discovered penicillin in Wright's own department. Reporters immediately besieged St. Mary's in search of Fleming, and interviews with him appeared in the press. Not to be outdone, Sir Robert Robinson, professor of chemistry at Oxford, followed with a letter to *The Times* suggesting that Florey's team deserved at least a bouquet. This drew a troupe of reporters to Florey, but he refused to see them and even forbade any member of his staff to talk to the press. To Americans this must seem unbelievable, but in Britain at that time scientists seeking "cheap publicity" in the daily press were thought by their colleagues to debase themselves and their lofty profession. The journalists therefore returned to Fleming, who told them about his own earlier work and also about the achievements of the Oxford team. They romanticized the former and ignored the latter, so much so that when Fleming, Florey, and Chain shared the 1945 Nobel Prize in physiology and medicine, only Fleming made the headlines, while mentions of Florey and Chain appeared in small print.

Fleming spent the remaining ten years of his life collecting twenty-five honorary degrees, twenty-six medals, eighteen prizes, thirteen decorations, the freedom of fifteen cities, and honorary membership in eighty-nine scientific academies and societies. A friend of his told Macfarlane that he collected honors as a schoolboy collects stamps and was delighted at any rare acquisitions. Effusive admirers soon hailed him as the greatest scientific genius of all time, and he became the subject of several hero-worshiping biographies, including one by André Maurois, which both Hare and Macfarlane criticized as misleading.[7]

Macfarlane's last chapter gives his assessment of Fleming as a scientist, but even without that the book contains enough material for the reader to make his own. Peter Medawar wrote that "at every level of endeavour scientific research is a passionate undertaking, and the Promotion of Natural Knowledge depends above all else upon a sortie into what can be imagined, but is not yet known."[8] The eighteenth-century scientist Count Rumford wrote of himself, "The ardour of my mind is so ungovernable that every object that interests me engages my whole attention, and is pursued with a degree of indefatigable zeal which approaches to madness."[9] Florey was driven by such passion, but not Fleming, who worked like an office clerk from nine to five, at which hour punctually he repaired to his club, which had a rule that "no bicycles, dogs, or women will be permitted on the premises." He seldom worked for more than six hours a day and spent Saturday and Sunday at his country home. Nowhere in his work is there any sign of imagination, and neither Macfarlane's nor other biographies contain a single quotation that suggests originality or burning curiosity. Macfarlane writes that Fleming had few original ideas of his own and distrusted those of others. The one quality that made him rise above the routine hospital bacteriologist was his acute observation of anything unusual, derived, Macfarlane believes, from his youthful roaming of the wilds of Scotland.

It cannot be said of Fleming, as Lampedusa wrote of the Leopard, "that flattery slipped off him like water off leaves in a fountain," but he must have been aware of his own limitations, because he told a friend repeatedly that he had not really deserved the Nobel Prize, and his friend wrote to Macfarlane that he had to bite his lips not to agree with him. Fleming used to say of himself that he "just played with microbes," and this was literally true.

However, intellectual brilliance is not the only quality that counts. In an age when human values are often ignored, it is heartening to read that Fleming himself, and all the other actors in the penicillin drama, behaved most honorably. For example, in 1942, when an employee of Fleming's brother was critically ill with meningitis, Fleming appealed to Florey for penicillin. Florey brought it to London but warned Fleming that he had not yet tested injections into the cerebrospinal fluid in animals. Fleming decided to take the risk and cured the patient. He then passed his notes to Florey to include in Florey's publication on clinical trials, which Florey did with acknowledgment to Fleming. Not all scientists would have behaved so generously. I know that Chain was bitter about Fleming's fame and the lack of his own. Some students once told me that they had been to a lecture by Chain. "What did he talk about?" I asked. "How Fleming had not discovered penicillin," they replied. It was not Fleming's fault that the media ignored Chain.

Fleming was also a man of courage. During the blitz, when most of St. Mary's staff were evacuated from London, Fleming stayed at his post, even though his home was bombed twice. He was taciturn, never laughed, seldom even smiled, but Macfarlane writes that he had an undefinable quality that inspired affection and respect. His easygoing love of games made him a popular member of the Chelsea Arts Club, and his simplicity added to his appeal as a popular hero. Perhaps Medawar's picture of the scientist is incomplete: success in research is a haphazard business, and great discoveries are not always made by great thinkers. Some are made by skilled craftsmen, some by observant watchmen, and some even by prosaic people doing a regular job because they are paid for it. Perhaps the most important lesson that scientists can learn from Macfarlane's book is that the solutions to some of our great problems may be staring us in the face and that we may be too blind to see them.

When Macfarlane chose Florey as the subject of a biography, he found a colorful, complex, eloquent, and forceful man to portray. Fleming had none of these attributes, and whatever thoughts he may have had were left unexpressed. Such people do not make rewarding subjects unless one embellishes them, as earlier biographers had done. Macfarlane has painted an honest picture and has yet contrived to write an absorbing story; it is a piece of medical history, less about the man than about the subtle and sometimes ironic interplay of science, chance, and personalities that is the stuff of which discoveries are made.

DISCOVERER OF THE ATOMIC NUCLEUS

Review of *Rutherford: Simple Genius* by David Wilson
(London: Hodder & Stoughton, 1983)

Rutherford was one of my early heroes, and Wilson's biography of
this great and lovable man has enlivened and enlarged, rather than
debunked, my youthful image. Rutherford was the man who created
the atomic age: a farmer's boy from New Zealand whose brilliance
and herculean energy brought him the presidency of the Royal
Society, a peerage, and honors from all over the world. Wilson goes
a long way to trace the mental paths and the passionate curiosity
that led Rutherford to his great discoveries. He paints a picture of
a towering, boisterous, stunningly able, outgoing, cheerful, irascible,
good-natured, generous, and compassionate man who delights above
all in the pursuit of experimental physics and feels sorry "for the
poor chaps who haven't got labs to work in."

Supported by the legacy of Prince Albert's farsighted interest
in science, an 1851 Exhibition Scholarship, Rutherford arrived in
Cambridge in September 1895, only a few months before Röntgen's
discovery of X rays and Becquerel's of radioactivity ushered in a
revolution in physics. He found college men "very capable, especially in conversation. It's a pity so many of them fossilise." Rutherford soon realized that only by scientific success could he make
himself socially acceptable and financially viable. "If one gets a man

like J.J. [J. J. Thompson, then Cavendish Professor of Physics] to back one up," he wrote to his fiancée in New Zealand, "one is pretty safe to get any position." Little did he know Cambridge! Three years later, when his scholarship expired, he applied for the chair in physics at McGill University in Montreal, hoping "that this may make J.J. act in respect of getting me something to do in Cambridge. I will probably go in for a fellowship this year." In fact, nothing materialized, and most of Rutherford's great discoveries were made at Montreal, and later in Manchester. He returned to Cambridge in 1920 as J. J. Thompson's successor to the Cavendish chair, but only after protracted negotiations, because the university authorities considered the honor of a Cambridge professorship to be worth a substantial drop in salary. According to Rutherford, "The chief trouble is the literary fraternity which is of late becoming more and more persistent in principle against the greater share for scientific purposes." *Plus ça change* . . .

I first saw Rutherford in the autumn of 1937, at a seminar given by his friend Niels Bohr, the great Danish theoretician. Bohr proposed a liquid drop model of the atomic nucleus, which appeared to sweep away one of the problems Rutherford had been trying to solve since his arrival at Cambridge. If the nucleus is merely a fluid drop composed of protons and neutrons, then it has no fixed structure, and many of Rutherford's experiments designed to solve that structure had been pointless. Young scientists will be relieved to learn from this book that even Rutherford did some "damn silly" experiments at times. At the seminar I was overawed by the giants of physics and sat on one of the back benches, but another graduate student, Fred (now Sir Frederick) Dainton, overheard Rutherford saying to Bohr after the lecture, "If mass disappears, energy will appear." Here and in several other places Wilson demolishes the myth that Rutherford failed to foresee the possibility that nuclear physics might have practical applications. He died in 1937, a year before one of his former German pupils discovered uranium fission. He therefore did not live to witness the terrifying implications of the atomic age and maintained his buoyant faith in the humane value of physics to the end.

"This country is judged," he said in a speech in 1921, "not by the size of its exports or its fleet, but by its contribution to knowledge." He failed to foresee that shrinking exports would also throttle the funds needed for the pursuit of knowledge. I was surprised to learn that in 1915 it was accepted wisdom that the French invented, while the Germans and British turned their inventions to profit: many of my contemporaries who worked in the Cavendish Laboratory had to go to America in order to find someone interested in turning their inventions to profit. I wonder what killed the British spirit of enterprise.

Rutherford died before I had any chance of attending his lectures. After his death spare reprints of his scientific papers were laid out in the attic of the lab, and research students were allowed to help themselves. I still have those reprints and look to them as models of the way science should be done. The experimental results are reported lucidly and concisely, with a minimum of jargon and mathematics; every conceivable objection is excluded by experiment rather than by argument, leaving no possible loopholes in the conclusions. From those papers and from the atmosphere around the place I became imbued with Rutherford's values, which Wilson characterizes as loyalty to your laboratory, extreme devotion to hard experimental work, and strong aversion to speculation beyond what is justified by the experimental results. When Crick and Watson lounged around, arguing about problems for which there existed as yet no firm experimental data instead of getting down to the bench and doing experiments, I thought they were wasting their time. However, like Leonardo, they sometimes achieved most when they seemed to be working least, and their apparent idleness led them to solve the greatest of all biological problems, the structure of DNA. There is more than one way of doing good science.

Did Rutherford make his discoveries by the hypothetico-deductive method postulated by modern philosophers of science? Like Napoleon, who did not win his battles by any fixed strategy, Rutherford did not follow any one method. A favorite one, stressed by Wilson, consisted of pursuing any anomalous or unexpected effect, but any intelligent scientist does that. Rutherford's strength lay in always being on the lookout for such effects and being extremely observant in spotting them. The most spectacular of those unexpected effects was obtained in 1909 by two of Rutherford's collaborators, Hans Geiger and Ernest Marsden, when they watched the scattering of alpha particles by a gold foil (alpha particles are helium nuclei shot out by their radium source). Most of the particles passed straight through the gold, but about one in eight thousand was reflected backward: in Rutherford's words, "as though you had fired a 15-inch shell at a piece of tissue paper and it had bounced back at you." This observation provided the first clue to the structure of the atom. It showed that it is a solar system in which nearly all the weight is concentrated in a tiny positively charged sun, the nucleus; this nucleus is surrounded by almost weightless negatively charged planets, the electrons. An alpha particle is itself a tiny atomic nucleus, and a gold foil therefore looks to it like empty space, in which the heavy gold nuclei are distributed so thinly that the chance of hitting one is quite small, but if the much lighter alpha particle does hit one of them, it is bounced back hard by its positive charge.

Geiger and Marsden's experiment had been designed to tell them something about alpha particles rather than about the gold foil, and Rutherford had had no working hypothesis of the atom before their totally unexpected result. I once held this up to Sir Karl Popper as an argument against the hypothetico-deductive method, which postulates that scientists advance by first formulating hypotheses and then designing experiments to test them, rather than the inductive method, which consists of deriving theories from observations. Popper retorted that neither Geiger nor Marsden had been able to derive the structure of the atom from their observations. Therefore, it was not implicit in them but was the brainchild of Rutherford's powerful physical insight. I have now learned that even for Rutherford the truth did not dawn in a flash; it took him eighteen months to work it out, which shows that he needed more than the bare observations. On another occasion in Cambridge many years later, two of Rutherford's collaborators, Mark Oliphant and John Cockcroft, obtained a result that neither they nor Rutherford could at first understand. It kept Rutherford awake. He turned it over and over in his mind, until at 3:00 A.M. he suddenly knew the answer. Excitedly he telephoned Oliphant to wake him up and tell him what it was. Here again Rutherford's imagination solved the riddle: they had discovered a light isotope of helium, helium 3, which has since found an important use in work at temperatures near absolute zero.

Wilson's most vivid descriptions of Rutherford's way of working are drawn from one of the very few among his collaborators who is still alive, the Russian physicist Peter Kapitsa.

Many admire Rutherford's intuition which told him how to set up the experiment and what to look for. . . . Intuition is usually defined as an instinctive process of the mind, something inexplicable which subconsciously leads to the correct solution.

This may be partly true, but is strongly exaggerated. The ordinary reader is simply unaware of the colossal amount of work done by scientists. . . . Anyone who has closely observed Rutherford can testify to the enormous amount of work he did. He worked incessantly, always in search of something new. He reported or published only those works which had a positive result; these, however, constituted barely a few per cent of the whole mass of work he did; the rest remained unpublished, and unknown even to his students.

Wilson mentions that Rutherford's failures sometimes drove him to black despair.

Rutherford's 1909 model of the atom failed to explain why the electrons do not fall into the nucleus to neutralize its charge. Bohr took up this problem. He combined Rutherford's model with Planck's quantum theory by postulating that, for reasons not then understood, the electrons keep in fixed orbits, and he demonstrated the validity of his theory by calculating correctly, for the first time, the wavelengths of the spectral lines given by the simplest atom, hydrogen. Rutherford's happy collaboration with Bohr contrasts strangely with his disdain for theoreticians. "Even the best mathematicians have a tendency to treat physics as purely a matter of equations," he said in 1907. "I think this is shown by the poverty of the theoretical communications on the problems that face the experimenter today." Coming two years after the publication of Albert Einstein's epoch-making papers on the photoelectric effect and relativity, this is an oddly blindered outburst. The French theoretician Louis Victor de Broglie reciprocated by condemning Rutherford's "concrete, simplistic modelling of the nucleus" and pointed out that "the fundamental laws of atomic behaviour can only be expressed in abstract terms." Physicists today would agree with de Broglie, but Wilson shows that Rutherford won, as always, because it was he and not the theoreticians who first predicted the existence of the neutron.

Wilson's book also shows the strange contrast between Rutherford's often narrow-minded and provincial utterances and his farsighted, generous, and internationalist actions. When Arthur Eddington verified the deflection of light by gravity predicted by Einstein's theory of relativity, Rutherford grumbled that this result might "draw scientific men away from experiment towards broad metaphysical conceptions." By contrast, when the Nazis dismissed

Max Born, the pioneer of quantum mechanics, from his chair in theoretical physics at Göttingen, Rutherford immediately moved heaven and earth to find him support, a niche to work in at the Cavendish Laboratory and a house in Cambridge. Rutherford disapproved of people who spoke several languages—"you can express yourself well in one language and that should be English"— yet he was in correspondence with all the prominent European physicists and went out of his way to help whoever was in professional or personal trouble. He helped Marie Curie when her love affair with the physicist Paul Langevin was picked up as a scandal by the French papers.

Far from being the jingoist his utterances imply, he maintained contact with his friend Stefan Meyer, the head of the Vienna Radium Institute, even when the outbreak of the First World War had made Meyer technically an enemy. In 1921, when the survival of Meyer's institute was threatened by runaway inflation, Rutherford saved it by getting the Royal Society to pay Meyer £500 for radium loaned to his Manchester laboratory before the war. When Jewish scholars were dismissed from German universities by the Nazis, Rutherford took a lead in founding the Academic Assistance Council, which raised money for their support and reestablishment in Great Britain. But strangely, for such an outspoken man, he refrained from any public condemnation of the Nazis' anti-Semitic policies for fear of offending the Germans and in a desire to remain "non-political." This timid attitude reflects a widespread misjudgment, which bolstered the Nazis' conviction that the British would never fight.

Rutherford's leadership at the Cavendish Laboratory was crowned by the remarkable achievements of 1932, when James Chadwick discovered the neutron, John Cockcroft and Ernest Walton split the lithium atom, and Patrick Blackett demonstrated the existence of the positron. These triumphs overshadowed the dawn of a new science in a small subdepartment of the Cavendish, the Crystallographic Laboratory. There J. D. Bernal, a colorful and brilliant young Irishman, was beginning to apply physics to the study of living molecules, such as proteins and viruses. As Bernal's research student, I was disappointed that Rutherford never looked in to find out what we were doing, and I thought this was because he was indifferent to sciences other than atomic physics. But Wilson relates that the conservative and puritanical Rutherford detested the undisciplined Bernal, who was a Communist and a woman chaser and let his scientific imagination run wild. I was surprised to read that

Rutherford had actually wanted to throw Bernal out of the Cavendish but was restrained from doing so by W. L. Bragg, Rutherford's successor at Manchester and at Cambridge. Had Bragg not intervened, Bernal's pioneering work in molecular biology would not have started, John Kendrew and I would not have solved the structure of proteins, and Watson and Crick would never have met.

I was disappointed to find the famous laboratory poorly equipped and some of its members making a virtue of necessity by boasting of the great discoveries that had been made with no more apparatus than string and sealing wax. Apparently Rutherford, though irritated by the chronic breaking down of machines, was uninterested in raising money for more reliable components. Wilson believes that despite his Jovian confidence in his scientific genius he lacked the confidence to ask for big money, but this seems doubtful—perhaps it was just not his style of working. To the detriment of the laboratory, Rutherford's failing was shared by Bragg, even though in his youth Bragg himself had suffered under the penny-pinching tyranny of the formidable Lincoln, the mustached laboratory steward who survived to my day.

In mitigation it must be said that Rutherford and Bragg were not interested in money for themselves either. Like Faraday, Rutherford never took out a patent and would have strongly disapproved of the gene technologists' present scramble for money. Rutherford was also averse to snobbery, even though he proudly accepted a peerage. But Wilson is wrong in describing his laboratory as free from class distinctions. There may have been none among the scientists, but there was a sharp division between them and the technicians, symbolized by separate tearooms. This aroused much ill-feeling when technicians came to include qualified engineers. Rutherford forbade people to work in the laboratory after 6:00 P.M., on the ground that they should go home and think, and for many years after his death, I was refused a key to the Cavendish site. If I wanted to switch off my X-ray tube at night, I had to climb over the tall wrought-iron gate and brave the wrath of Alf the porter patrolling the dimly lit courts.

None of this mattered very much. What did matter Kapitsa described to his mother when he arrived at the Cavendish in 1921.

With us in Russia everything was cut according to the German pattern. . . . But England provided the most outstanding physicists and I now begin to understand why: the English school

develops individuality . . . and provides infinite room for man-
ifestation of personality. . . . Here they often do work which is
so incredible in its conception that it would simply be ridiculed
in Russia. When I asked why . . . it emerged they were simply
ideas of young people, but the Crocodile [Kapitsa's nickname
for Rutherford] values so highly that a person should express
himself that he not only allows them to work on their own
themes, but also encourages them and tries to put sense into
their sometimes futile plans. The second factor is the urge to
achieve results.

This was still true when I left the Cavendish Laboratory forty-one
years later.

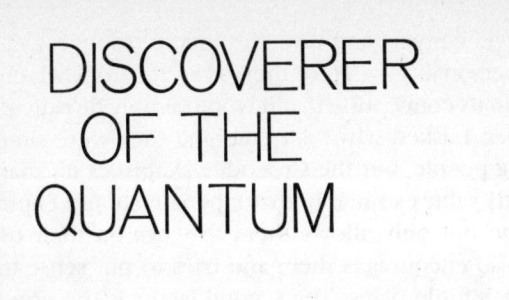

DISCOVERER
OF THE
QUANTUM

Review of *The Dilemmas of an Upright Man: Max Planck as a Spokesman for German Science* by J. L. Heilbron (Berkeley: University of California Press, 1986)

The dilemmas referred to in the title of this book were those faced by a leading German scientist who believed in his country right or wrong, even when that country became the embodiment of evil. Max Planck was a physicist, famous to this day for his introduction of the quantum theory. He was born in 1858 in Kiel, which was then part of Denmark. One of his formative memories was the triumphant entry in 1864 of Bismarck's Prussian troops, which united that German province with Prussia. His elder brother's death in the Battle of Verdun, during the Franco-Prussian War of 1870–71, "made him feel at one with the heroes who sealed their true love for the fatherland with their own blood." These were noble sentiments in those days. At school in Munich, where his father had been made a professor of law, he nearly always earned the annual prize for religion and good behavior. His teachers described him as conscientious, open, cheerful, gifted in all subjects, especially mathematics, yet modest and popular with his classmates. He was also intensely musical and had absolute pitch. He wondered whether to study classics, music, or physics and finally opted for physics, even though a leading physicist advised him that there was nothing significant left to be discovered in that subject. Planck found nothing to rebel against until he was over forty, when the dogged pursuit of a vital physical problem led him, almost against his will, to make a revolutionary discovery.

Until then he wrote treatises on the physics of heat, first as a *Privatdozent* (unpaid) in Munich, later as a professor in Kiel, and finally in Berlin. They aroused little interest. His appointment at Kiel was given "confident that he would remain faithful in unbreakable loyalty to His Majesty the Emperor and to the Imperial Family." To Planck that read not as an empty phrase but as a sacred duty, to which he still felt bound thirty-three years later, after the collapse of the German armies in October 1918. He then wrote to Einstein,

> It would be a great stroke of fortune for us, indeed a saving grace, if the bearer of the crown would voluntarily renounce his rights. But the word "Voluntary" makes it impossible for me to come forward in the matter; for first I think of my sworn oath, and second, I feel something that you will not understand at all, . . . namely, piety and an unbreakable attachment to the state to which I belong and which is embodied in the person of the monarch.

Two days afterward the Reichstag declared a parliamentary republic that threw Germany into chaos.

Planck's epoch-making discovery came in 1900 and dealt with the interaction between radiation and matter. The question that exercised German physicists concerned the color and intensity of radiation emitted by a hot black body. The experimentalists had developed sensitive methods for measuring the radiation, and a theoretician, Wilhelm Wien, had derived an apparently well-founded mathematical theory to account for their observations. However, as techniques were refined and temperatures raised, deviations from Wien's formula became apparent. Planck therefore modified Wien's formula in a way that fitted the observations exactly. A lesser man would have been content with that, but Planck had already been struggling unsuccessfully for six years with the formulation of a fundamental law for the interaction between radiation and matter; he felt that "his formula had only limited value," since even if it proved accurate, it was only "happily guessed." "Therefore," he wrote, "from the day of its formulation, I was occupied with the problem of obtaining for it a true physical meaning. . . . After several weeks of the most strenuous work of my life the darkness lifted, and a new unexpected perspective began to dawn on me."

Classical physicists, including Planck, were skeptical of the Austrian physicist Ludwig Boltzmann's interpretation of heat in terms of atomic vibrations and preferred to think of matter as a continuum. Yet it soon became clear to Planck that his radiation law conflicted with that picture. Many years later Planck confessed to the American physicist R. W. Wood, "What I did can be summarized simply as an act of desperation. . . . The two laws of thermodynamics . . . must be upheld under all circumstances. For the rest I was prepared to sacrifice every one of my previous convictions about physical laws." Both Wien and Planck had thought of the black body radiation as being emitted by small oscillations of electric charges in that body, but neither of them had regarded the oscillators as atoms. In his desperation Planck decided to use Boltzmann's atomistic approach to formulate the equilibrium between radiation and matter. The results of his mathematical work forced on him the strange conclusion that the energy available to the oscillators was not continuous "but composed of a definite number of finite equal parts . . . or energy elements." Planck called one such element the quantum of action; it soon became known as Planck's constant.

His concept was so revolutionary that no one, not even Planck himself, immediately realized its implications. One man did, a few years later. That man was Einstein. Planck meanwhile was more concerned with another universal constant that had emerged in his new radiation law. He named it in honor of Boltzmann, to whom he owed his successful attack on the problem, and he used it to calculate accurately for the first time the true weights of single atoms and the exact charge carried by the recently discovered electron. Boltzmann's constant was derived from nothing more than the laws of heat; yet when taken together with Newton's gravitational constant and the speed of light, it provided a system of natural units of mass, electric charge, energy, and time independent of human observers and valid throughout the universe. That discovery convinced Planck that there exists a physical universe independent of our senses.

Planck's insistence on the reality of atoms that no one had yet seen, and on the universal validity of physical constants derived from the laws of heat, brought him into sharp conflict with the Positivist philosophers. The author of the book under review assumes that his readers know what Positivism was all about, but I did not and looked it up in the *Encyclopaedia Britannica*. I learned that its founder, the nineteenth-century French philosopher Auguste Comte, called his *Cours de Philosophie* "Positive" because it was concerned only with positive facts. According to Comte, the sciences have to study the facts and regularities of nature and formulate them as descriptive laws and not, as Planck had done, interpret their meaning in terms of a reality that cannot be observed directly. The twentieth-century Viennese school of Positivists distinguished between a metaphysical utterance and a genuine proposition by requiring the latter to be conclusively verifiable, which neither atoms nor quanta then were. Planck's chief opponent, the philosopher-physicist Ernst Mach, went so far as to argue that all factual knowledge consists of conceptual organization and elaboration of immediate experience; he therefore denied the existence of Kant's "things in themselves," the ultimate entities underlying phenomena.

For the first time the peaceful and good-natured Planck turned into a fighter. He vigorously asserted the reality of the unseen world that had emerged from his mathematical work and declared that "the philosopher who limits himself to asking to what extent the meaning of a new idea is evident a priori hampers the development of science. What matters is that the idea gives rise to useful work. A positivism that rejects every transcendental idea is as one-sided as a metaphysics which scorns individual experience." Mach replied disdainfully, "If belief in the reality of atoms is so essential to you, I will have no more truck with physical thinking." Many years later Planck complained, "I have never succeeded in getting everyone to agree to a new result, the truth of which I could demonstrate by a conclusive, but purely theoretical argument. That has been one of my most painful experiences."

When Planck derived his radiation formula, he parceled into quanta only the energy of the oscillators in the black body and left the physical meaning of his quantum of action unclear. Five years later Einstein published his famous trio of papers, among them "On a Heuristic Viewpoint Concerning the Production and Transformation of Light," which extended and deepened Planck's work on the interaction between radiation and matter. Einstein opened his paper with the revolutionary proposal that radiation (in other words, light itself) consists of quanta. An ingenious mathematical argument then led him to conclude that these quanta were the same as Planck's. Einstein showed that his theory points to an understanding of an enigmatic effect discovered, ironically, by Heinrich Hertz, the very man who was thought to have proved that light consists of electromagnetic waves. This is the photoelectric effect, so called because light falling on certain metals, such as selenium, causes an emission of electrons. Einstein showed that the effect is caused by a quantum of light absorbed by a metal atom transferring its energy to, and liberating, an electron. His paper implied that all energy is parceled into quanta, whether it is contained in radiation or atoms. This was the true meaning of Planck's radiation law.

Planck was deeply impressed by Einstein's papers. A meaner character might have feared to be overshadowed; Planck engineered the establishment of a research professorship free of any duties for Einstein in Berlin. Their admiration was mutual; Einstein, who was no respecter of persons, later wrote that the best thing about Berlin was the sheer joy of being near Planck, even though the two men's views differed on almost any subject other than science. Planck was revered by all who knew him. He was honest, modest, and gentle, entirely free of the pomp of the proverbial German professor. Lise Meitner, who had worked with him for forty years, wrote that he never decided anything on the ground that it might be either useful or harmful to himself but always on the true merits of the case. Contrast this with John Keynes's dictum about Lloyd George: "He made every decision on grounds other than the real merits of the case."

In fact Planck was something of a saint. He wrote that he was deeply religious even though he did not believe in a personal, let alone a Christian, God. Planck's beliefs are well expressed in his published lectures on the philosophy of physics, in which he wrote, "There is a real world independent of our senses; the laws of nature were not invented by man, but forced upon him by that natural world. They are the expression of a rational world order. Both religion and natural science need a belief in God, but to the former God is the starting point, to the latter the goal of every train of thought." Elsewhere he wrote: "Physics must contain no contradictions; in terms of ethics this comes down to honesty and truthfulness. Since justice is inseparable from truthfulness, it must be the same for all, just as the laws of Nature." This comes close to Jacques Monod's later attempt to build an ethic upon scientific truth. While Planck was the grandson of a Protestant pastor and Einstein was the son of a Jewish businessman, both believed in a rational, God-made order of the world, independent of man. Both men also believed in determinism and absolute causality, and were therefore equally reluctant to accept Erwin Schrödinger and Max Born's probabilistic wave mechanics as the ultimate picture of the atomic world.

I was sad to read that Planck's faith and goodness did not prevent him from being infected by Germany's chauvinistic fever at the start of the First World War, probably because devotion to the glory of the fatherland reached an intensity that went far beyond people's normal love for their native land and found its focus in the person of the emperor. Certain of the righteousness of the German cause and of the civilized nature of all German people, Planck indignantly rejected as lying propaganda Allied reports of German atrocities in Belgium until, two years later, his Dutch physicist friend Hendrik Lorentz convinced him otherwise.

Planck's naïve faith in German goodness was to be shaken further soon after the war by vicious attacks on Einstein. I had been under the illusion that the excesses of anti-Semitism started only after Hitler's accession to power in 1933, but I learned from Heilbron's book that attacks on Einstein began in 1919, the very year Sir Arthur Eddington's confirmation of the theory of relativity brought him world fame. The attacks came from anti-Semitic physicists who rejected Einstein's theories, from anti-Semitic newspapers, and from Nazi thugs who actually threatened to kill him, a pacifist who had done no one any harm and whom Planck regarded as one of the greatest physicists of all time. Planck had stood up against official anti-Semitism on two earlier occasions: in 1894, when the Prussian minister of education tried to evade the science faculty's recommendation to appoint a Jew, Emil Warburg, to the chair of experimental physics in Berlin, and in 1895, when the ministry demanded the dismissal of a Jewish *Privatdozent* whose only offense was his support of the Social Democratic party. Planck responded to the attacks on Einstein not by publicly denouncing the anti-Semites but by arranging a public discussion between Einstein and the chief antirelativist, Philipp Lenard, in which "opposing conceptions of the epistemological foundations of science confronted each other in a dignified manner," but that did nothing to silence the slanderous campaign.

In 1933, when Hitler took power, Planck, as the doyen of German science, was secretary of the Prussian Academy of Sciences and president of the Kaiser Wilhelm Gessellschaft, the forerunner of the present Max-Planck Gessellschaft, which ran Germany's independent research institutes, including Einstein's in Berlin. Einstein happened to be in the United States. Having learned of the dismissal of Jewish scientists, Einstein publicly declared that he would not return to Germany because it no longer recognized "civil liberty, tolerance and equality of citizens before the law." The Nazi press responded to Einstein's statement with a flood of abuse, and the Reichskommissar appointed by the Prussian minister of education to take charge of the academy demanded the institution of a disciplinary procedure against Einstein. Planck realized the hopelessness of a compromise and advised Einstein to resign "so as to conserve his honourable relations with the Academy and save his friends an immeasurable amount of grief and pain."

Planck had no understanding for Einstein's public protests, believing that as a German Einstein should have stood up for Germany abroad whatever the faults of its new regime. He told Einstein that his public statements had given all his friends great pain. "There has taken place a collision between two conflicting views of the world. I have no understanding for either. I feel remote from yours, as you will remember from our talks about your propaganda for refusing military service" (during the First World War). At an official meeting of the academy, Planck declared that its members had a special duty of loyalty to the government and regretted that Einstein's political attitude had made his continued membership impossible. For the record, he stressed that Einstein was not just an eminent physicist but *the* physicist, through whose papers, published by the academy, physical knowledge in our century had gained in depth to an extent comparable only with that achieved by Kepler and Newton. I find it hard to see how Planck could have expected Einstein to be loyal to a government composed of men who had long treated him as a criminal. Einstein never forgave Planck for what he regarded as his cowardly failure to stand up for him and other Jewish colleagues.

Fritz Haber, the Jewish chemist whose synthesis of ammonia from nitrogen of the air had saved the German army from running out of explosives soon after the start of the First World War, had committed no indiscretions, but he was dismissed all the same. Planck sought an audience with Hitler to get Haber reinstated. When he extolled Haber's and other Jewish Germans' contributions to science, Hitler replied that he had nothing against Jews as such, but that they were all Communists. When Planck tried to argue, Hitler shouted, "People say that I get attacks of nervous weakness, but I have nerves of steel"; he slapped his knee and whipped himself into a rage that continued until Planck took his leave. Planck told Born, another eminent Jewish physicist dismissed by the Nazis, that this interview extinguished any hopes he might have had of openly exerting his influence in favor of his Jewish colleagues. He did so on only one occasion, two years later, after Haber had died in exile, shunned as a Jew by the Germans and as the initiator of Germany's gas war by his British and French colleagues. Planck decided that the Kaiser Wilhelm Gesellschaft should hold a meeting in Haber's memory. When the invitations had been sent, the minister of culture forbade all public employees to attend. Threats notwithstanding, Planck said that he would hold the meeting unless the police dragged him out. Another state-employed physicist, the Nobel laureate Max von Laue, attended despite the minister's veto, but he suffered a heart attack afterward, symptomatic of the tensions in which Germans lived at the time.

When Laue went on a lecture tour to the United States, Planck asked him to make people understand the difficulties he had to face and to assure them that "quieter and more normal times would return." This had been Planck's experience during the revolution, civil war, and inflation after the First World War, when his unique standing had made it possible for him to rebuild and even to strengthen the fabric of German physics despite the turbulent times. Planck felt confident of being able to do the same again; meanwhile he tried discreetly to minimize the damage by keeping administrative power in his own hands rather than letting it be seized by the Nazis. He therefore refrained from public protests, tried to stop the dismissal of Jewish heads of Kaiser Wilhelm institutes by quiet diplomacy, and dissuaded German scientists from emigrating abroad. One of them was Werner Heisenberg, the founder of quantum mechanics, who had been the subject of vicious attacks by the Nazi press for his support of Einstein's theory of relativity. Heisenberg stayed and later headed the German atomic bomb project. Fortunately this proved unsuccessful. Planck's failure to stand up in public against the atrocities of the Nazi regime and his discreet maneuvers to save German science were a disaster, because they corroborated Nazi propaganda that branded reports about the real goings-on in Germany as Jewish inventions.

Planck's nineteenth-century ideal of "a clear conscience that expresses itself in conscientious fulfilment of one's duty" proved an insufficient guide in those evil times. His hopes for better days were shattered by the final, terrible tragedy of his life. His eldest son, Karl, had been killed in the First World War, and both his daughters had died giving birth to their first children. The only child left was his son Erwin, who was also his closest friend. After the attempt on Hitler's life in July 1944, Erwin, who had been a high official in the ministry of defense in the Weimar Republic, was arrested. According to historians who unearthed the police evidence against him, Erwin had discussed possible ways of overthrowing the Nazi regime with several groups from 1934 onward and had known some of the conspirators of the July plot but had not been aware of the actual plot. Since father and son were very close, Heilbron believes that the father must have known of these activities. Perhaps this partly explains Planck's public reticence and his confidence in better days to come. For several months after Erwin's arrest, Planck was torn between hope and despair, until he received the shattering news that the People's Court had sentenced Erwin to death. Planck then wrote to both Hitler and Himmler assuring them that his son had had no knowledge of the plot, and in early February 1945 he was told of an imminent reprieve. Five days later Erwin was hanged. In a letter which conveyed that news to his old friend the physicist Arnold Sommerfeld, Planck wrote, "My sorrow cannot be put into words. I struggle for the strength to give meaning to my future life by conscientious work." Even after that final tragedy, this characteristically German ideal remained his guiding star.

DISCOVERERS OF THE DOUBLE HELIX

From *The Daily Telegraph* [London], 27 April 1987

One day in 1950 a strange young head with a crew cut and bulging eyes popped through my door and asked, without saying so much as hello, "Can I come and work here?" He was Jim Watson, who wanted to join the small team of enthusiasts for molecular biology at the Cavendish Laboratory in Cambridge, which I led.

My colleagues were John Kendrew, a chemist like myself, and Francis Crick and Hugh Huxley, both physicists. We shared the belief that the nature of life could be understood only by getting to know the atomic structure of living matter, and that physics and chemistry would open the way, if only we could find it.

In his best-selling book *The Double Helix*, Watson mirrors himself as a brash western cowboy entering our genteel circle, but this is a caricature. Watson's arrival had an electrifying effect on us because he made us look at our problems from the genetic point of view. He asked not just What is the atomic structure of living matter? but, foremost, What is the structure of the gene that determines it? He found an echo in Crick, who had begun to think along similar lines. Crick was thirty-four, a more than mature graduate student because of years lost by the war; Watson was twenty-two, a whiz kid from Chicago who had entered college at the age of fifteen and got his Ph.D. in genetics at twenty.

They shared the sublime arrogance of men who had rarely met their intellectual equals. Crick was tall, fair, dandyishly dressed; he talked volubly, each phrase of his King's English strongly accented and punctuated by eruptions of Jovian laughter that reverberated through the laboratory. To emphasize the contrast, Watson went around like a tramp, making a show of not cleaning his one pair of shoes for an entire term (an eccentricity in those days), and dropped his sporadic nasal utterances in a low monotone that faded before the end of each sentence and was followed by a snort.

To say that they did not suffer fools gladly would be an understatement: Crick's comments would hit out like daggers at non sequiturs, and Watson demonstratively unfolded his newspaper at seminars that bored him. Watson had put Crick's mind to the structure of DNA, yet their relationship resembled that of teacher and pupil because there was little science that Watson could teach Crick but much that Crick could teach Watson. Crick had a profound understanding of that hardest of the sciences, physics, without which the structure of DNA would never have been solved. This crucial fact is obscured in Watson's *Double Helix*. Yet Watson had an intuitive knowledge of the features that DNA ought to have if it were to make genetic sense.

At one stage there was much argument about whether genes consist of two or three chains of DNA wound around each other. Watson took French lessons with a lady who kept a boardinghouse for young girls wanting to learn French. One day she noticed him pacing restlessly and muttering to himself, "There must be two ... there must be two. . . ." She guessed that this referred to matters of the heart. But we knew better. He reasoned on genetic grounds that genes must be made of two chains of DNA, and he was right.

Like Leonardo, Crick and Watson often achieved most when they seemed to be working least. They did an immense amount of hard studying hidden away, often at night, but when you saw them they were more likely to be engaged in argument and apparently idle. This was their way of attacking a problem that could be solved only by a tremendous leap of the imagination, supported by profound knowledge. Imagination comes first in both artistic and scientific creations, but in science there is only one answer and that has to be correct.

In his *Double Helix*, Watson makes Rosalind Franklin, who was simultaneously trying to solve DNA at King's College, London, into an aggressive, blinkered bluestocking. When I saw Watson's draft I was furious about his maligning that gifted girl, who could not defend herself because she had died of cancer in 1958. Rosalind came from a family of London bankers who expected her to marry and be a society lady rather than waste her time on abstruse scientific problems. Not that she was unattractive or did not care about her looks. She dressed much more tastefully than the average Cambridge undergraduate. But her family's opposition generated in her a fierce, uncompromising ambition to prove herself as a scientist.

She tried to solve DNA slowly, methodically, on the basis of her experimental results, and she resented Watson and her London colleague Maurice Wilkins's breathing down her neck. She respected Crick but rejected his idea that DNA forms a helix, even though some of her results strongly hinted that it does. Instead, she got stuck in a blind alley. Her notes show that she was just trying to extricate herself when Watson and Crick solved the problem. Given time, she might have found the right answer, and had she lived she would have been a strong candidate for a share in the Nobel Prize.

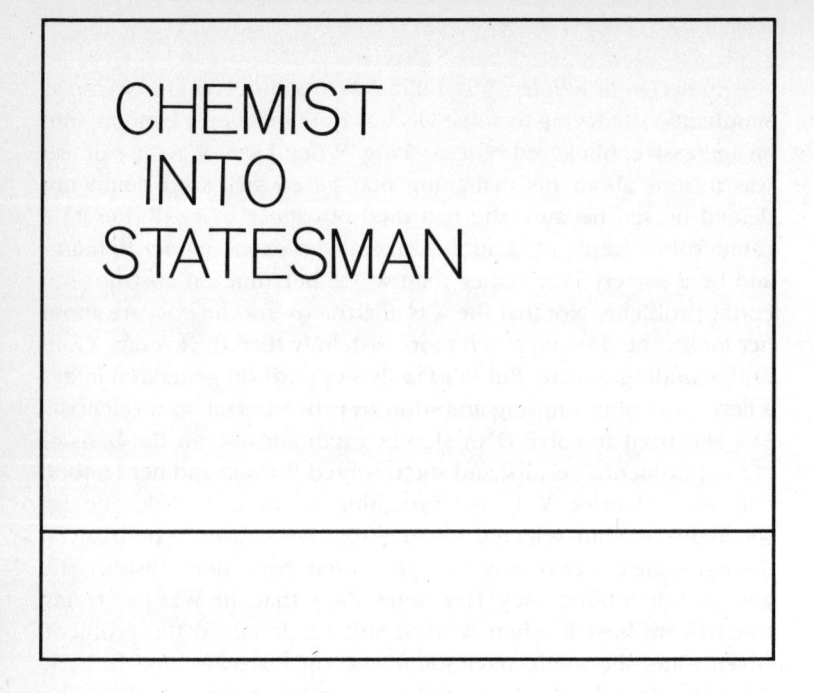

CHEMIST
INTO
STATESMAN

Chaim Weizmann was born in 1874, the son of a timber merchant, in the Pale of Settlement, "that prison house created by Tsarist Russia for the largest part of its Jewish population."[1] Outside the pale, Jews were not even allowed to register in a hotel, and anyone who put up a Jew privately without reporting him to the police was himself subject to arrest. Admission to universities was subject to the *numerus clausus* that restricted the proportion of Jewish students to the proportion of Jews in the Russian population. Jews were therefore given a special set of difficult papers in their entrance examination, and even if they passed were likely to be refused a residence permit in a university town. (Things were not all that different under Brezhnev's regime: one of my Russian colleagues, a physical chemist, was ordered to fail a Ph.D. student's thesis because he was Jewish.) To escape from these restrictions, the eighteen-year-old Weizmann took a job as a bargeman on a timber raft going down the Vistula, made off as soon as it had crossed the German frontier, and enrolled as a chemistry student at the University of Darmstadt (outside Russia and Turkey, passports were not needed for travel in Europe before 1914). After two years at Darmstadt, Weizmann studied for a year in Berlin, where he recalls nostalgically that students could get theater tickets for fifty pfennig and opera tickets for one mark. He then followed his Berlin professor to Fribourg in Switzerland, where he got his Ph.D. in 1899 for a thesis on "the electrolytic reduction of 1-nitroanthrachinone and the condensation of phenanthrachinone with 1-nitroanthrachinone." These were compounds related to dyestuffs, whose synthesis earned Weizmann some much-needed money.

From the moment he left Russia as a student, Weizmann was determined to raise the Jews to a nation equal in rank with other nations. Throughout his student days he divided his time between chemistry and Zionist propaganda, carried on openly among his fellow Russian students in Germany and Switzerland and clandestinely among the people in the pale when he returned to Russia during vacations. In Switzerland his proselytizing brought him into conflict with Lenin and Trotsky, who had made Swiss universities their recruiting ground for young revolutionaries, and who were intolerant of expressions of a separate Jewish identity. Weizmann's early quarrels with Lenin sowed the seed for the conflict between Bolshevism and Zionism that continues to this day. Ironically, the two men looked so much alike that on one occasion Weizmann was followed by police detectives who mistook him for Lenin.

Weizmann's Ph.D. thesis earned him the title of *Privatdozent* with the right to lecture at the University of Geneva—alas, without a stipend. This proved no obstacle, because a patent on a novel synthesis of a dyestuff brought him a good salary from the German chemical giant I. G. Farbenindustrie. Although Weizmann was highly successful at Geneva, he looked on England as the country where a Jew might work without hindrance and be judged entirely on his merits. After four years, now aged thirty, he therefore decided to make a fresh start in the chemical laboratory of the University of Manchester. Here he became the founder of what we now call biotechnology, which was to enable him to become the founder of the state of Israel. He foresaw the need for a substitute for natural rubber and therefore looked around for a starting material for its synthesis. Isoamylalcohol was such a material but would have been expensive to synthesize. Weizmann decided to try to let Nature do the job and used the university vacations to become apprenticed in bacteriology at the Pasteur Institute in Paris, where he set out to find a microorganism that would produce isoamylalcohol by fermentation from sugars. He discovered a bacterium, since named *Clostridium acetobutyricum Weizmann*, that did indeed ferment sugars to a liquid that smelled like isoamylalcohol but on analysis turned out to be a mixture of isobutylalcohol and acetone. Weizmann neither published nor patented this result until much later, because it never occurred to him that it might prove important.

After the outbreak of war in 1914, England found itself acutely short of acetone, which was needed as a solvent for the explosive nitrocellulose. Weizmann had never made more than a few hundred milliliters by his fermentation process, but Winston Churchill, then first lord of the Admiralty, sent for him and demanded 60,000 metric tons of it. Weizmann turned chemical engineer, set up a pilot plant to make acetone from maize starch in a gin factory outside London, and then helped the Admiralty to build a factory in one of its arsenals and converted several distilleries to his process.

Lloyd George, in his *War Memoirs*, represented the Balfour Declaration as the British government's gift to the Jews in return for Weizmann's vital contribution to the war effort. Weizmann himself disputes this. He relates that his negotiations for a Jewish state in Palestine began as early as 1906, when he first met Arthur Balfour, then Conservative candidate for Parliament from Manchester. His account of the complex and protracted negotiations reminded me of Francis Bacon's dictum "In all negotiations of difficulty, a man may not look to sow and reap at once, but must prepare business and so ripen it by degrees," but that ripening would have borne no fruit, had it not been for the prestige and ready access to government ministers that Weizmann earned by his success as a chemist.

The declaration took the form of a letter from Balfour, then a cabinet minister, to Baron Edmond de Rothschild. Here is its text:

Dear Lord Rothschild,

I have much pleasure in conveying to you, on behalf of His Majesty's Government, the following declaration of sympathy with Jewish Zionist aspirations which has been submitted to, and approved by, the Cabinet.

His Majesty's Government view with favour the establishment in Palestine of a national home for the Jewish people, and will use their best endeavours to facilitate the achievement of this object, it being clearly understood that nothing shall be done which may prejudice the civil and religious rights of existing non-Jewish communities in Palestine, or the rights and political status enjoyed by Jews in any other country.

I should be grateful if you would bring this declaration to the knowledge of the Zionist Federation.

Signed: Arthur James Balfour.
2 November 1917

It is amazing that Weizmann, an immigrant Russian and self-appointed leader of the Jews, achieved this triumph single-handed. For much of the time he was at loggerheads with, rather than backed by, a bickering Zionist Congress. It seems that he was a born statesman, whose force of personality made other statesmen accept him on his own terms. According to Isaiah Berlin, "he possessed tact and charm to a degree exceeded by no statesman of modern days." He appealed:

> to reason rather than feeling. His method of argument was neither a demonstration founded on statistical or other carefully documented evidence, nor emotional rhetoric, nor a sermon addressed to the passions; it consisted in painting a very vivid, detailed, coherent, concrete picture of a given situation or course of events; and his interlocutors felt that this picture coincided with reality and conformed to their own experience of what men and events were like, of what had happened, or might happen, or could not happen; of what could and what could not be done. The moral, historical, economic, social and personal factors were blended in Weizmann's remarkable, unrecorded expositions much as they combine in life (this he spoke most effectively face to face, in private, and not before an audience). . . . He was prepared to conceal facts, to work in secret, to fascinate and enslave, individuals, to use his personal followers or anyone who appeared to him useful, as a means for limited ends — only to drop all interest in them, once the need for them was at an end.[2]

Something similar was said of Winston Churchill.

Weizmann's tremendous moral courage and supreme confidence in his own persuasive powers made him plead his case with political leaders as diverse as Lloyd George, Roosevelt, Mussolini, and ibn-Saud. Ben-Gurion wrote of him: "I never failed to be astonished by his inner forcefulness, by his determined manner. He could be angry with these men in power . . . but his anger emerged with such natural dignity that it was always deeply moving."³ Lord Robert Cecil wrote of his subdued enthusiasm and the extraordinary impressiveness of his attitude, "which make one forget his rather repellent and even sordid exterior."

Weizmann had a fervent belief that the Jews were a nation entitled to the rights of a nation; at the same time he was inspired by liberal ideals and averse to any form of violence. His high moral standards can be gauged from his courageous address to the Zionist Congress of 1945, at a time when the British Labour government was blocking Jewish immigration into Palestine and bitterness was at its height. "It is difficult in such circumstances to retain a belief in the victory of peaceful ideals, in the supremacy of moral values. And yet I affirm . . . that we must retain it. Zionism is a modern expression of the liberal ideal. Divorced from that ideal it loses all purpose, all hope. . . . Assassination, ambush, kidnapping, the murder of innocent men, are alien to the spirit of our movement. . . . [They mock] the ideals for which a Jewish society must stand." Already in 1929 Weizmann had written to Einstein, "I have always preached most unpopular realities to the Zionists, have always been attacked most bitterly." For example, he told the Congress in 1923, "Palestine is not an empty country . . . there are 500,000 Moslems, 100,000 Christians and a 100,000 Jews . . . there has been a striving on the part of the Arab people for a revival . . . being anxious for a revival of the scattered Jewish people, we treat with respect and reverence any attempt at revival amongst other people."⁴ In his autobiography, *Trial and Error*, Weizmann wrote, "We must stand by the ancient principle enunciated in our Torah: 'One law and one manner shall be for you and for the stranger that sojourneth with you.' *I am certain that the world will judge the Jewish State by what it does to the Arabs.*" There is never a word about the Jews being the chosen people. Weizmann regarded Menachem Begin as "a megalomaniac suffering from a Messianic complex."⁵

Although Weizmann was fond of quoting the prophets, he rarely appealed to God. I gained the impression that he was an agnostic at heart and believed in a secular Judaism as a cultural bond that welds the world's Jews into a single nation. His opposition to bigoted orthodoxy first emerges from a letter to Theodor Herzl in 1903. "If there is anything in Judaism that has become intolerable and incomprehensible to the best Jewish youth, it is the pressure

to equate its essence with the religious formalism of the orthodox."
In 1949 he wrote, "It is the new, secularized type of rabbi ... who
will make a heavy bid for power by parading his religious convictions ... they transgress a fundamental principle which has been
laid down by our sages: 'Thou shalt not make of the Torah a crown
to glory in, or a spade to dig with.' "[6] In today's Israel such people
have become a strong political force. Isaiah Berlin wrote, "Weizmann never called upon the Jews to make terrible sacrifices, or
offer their lives, or commit crimes, or condone the crimes of others,
for the sake of some felicity to be realized at some unspecified date
as the Marxists did. He wished to make his nation free and happy,
but not at the price of sinning against any human value in which
he believed."[7]

In his eagerness to awaken Jews all over the world to a national
consciousness, and to preserve Jewish culture and heritage in a
sovereign Jewish state, Weizmann was intolerant of those who had
become assimilated to the countries they lived in. "I don't care
what will happen to the Jewish communities in England or in
France ... their highest ideal is assimilation, disintegration, dissolution."[8] By contrast it seems to me that all the world's Jews could
not find refuge in Palestine even if they wanted to, and that by
failing to assimilate they will keep anti-Semitism alive forever. As
Boris Pasternak made Dr. Zhivago say, " 'Why don't the intellectual
leaders of the Jewish people ... say to the Jews: "Don't hold on to
your identity, don't all get together in a crowd. Be with all the
rest.... What the Gospels tell us is ... that there are no nations, only
people." ' "

In the Balfour Declaration the extent of the Jewish National
Home had been left open. Weizmann hoped that it would include
a much larger area than the present state of Israel. The Turkish
province of Palestine included both Cis and Transjordan, and at the
Versailles Peace Conference in 1919 Weizmann pleaded that all
territory west of the Hejaz railway, which runs from 'Aqaba via
Amman to Damascus, should be included in the National Home.
Today this may sound like imperialism, but judged by the noncha-
lance with which Lloyd George and Clemenceau disposed of the
former Turkish Empire by carving it up between them, Weizmann's
claim to the whole of Palestine seems less unreasonable now than,
say, France's to Syria and Lebanon. According to Asquith, Lloyd
George backed Weizmann, although he "does not care a damn for
the Jews, but thinks it an outrage to let the Holy places pass into

the possession of agnostic, atheistic France."[9] However, Isaiah Berlin believed that Asquith traduced Lloyd George, who did not like the English establishment and had a certain fellow feeling for minorities treated *de haut en bas* by grand Englishmen such as Lord Robert Cecil.

The opening, in 1925, of Hebrew University on Mount Scopus, for which Weizmann had campaigned since 1902, was one of his life's great triumphs. "Our Hebrew University . . . will have a centripetal force attracting all that is noblest in Jewry throughout the world."[10] Weizmann also labored for the other two academic institutions that have made Israel a great outpost of science and learning, the Weizmann Institute of Science and the Haifa Technion.

Weizmann's tragedy came when, in the critical years after 1938, he failed to persuade British leaders to open the gates of Palestine either to the Jews threatened by the holocaust or to its survivors. Torn between his passionate devotion to the Jewish cause and his admiration for Britain, the country that embodied his liberal ideals, he became estranged from both his English and his Jewish friends. General Smuts, another statesman with whom Weizmann made friends, wrote this epitaph: "He was the greatest Jew since Moses." I wish that the liberal ideals for which he stood would still guide Israel's rulers today.

ABOUT
SCIENCE

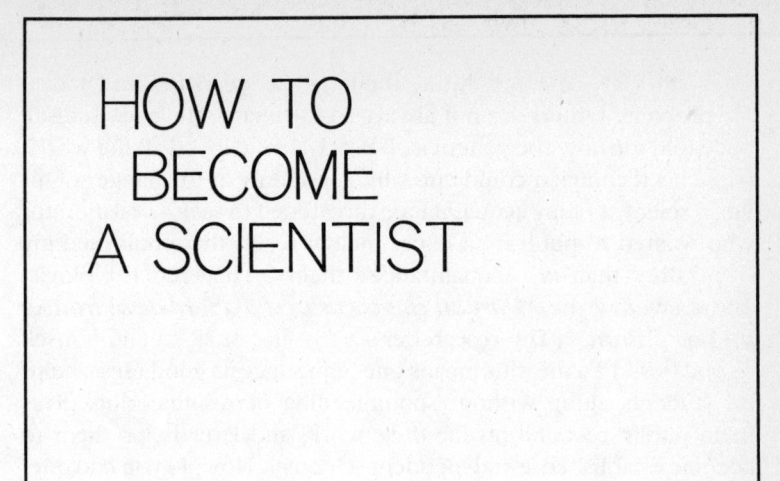

HOW TO
BECOME
A SCIENTIST

Review of *Advice to a Young Scientist*
by P. B. Medawar
(New York: Harper & Row, 1979)

This is a guidebook to the scientific scene, full of urbane wisdom, happy phrases, and entertaining examples. "How can I tell if I am cut out to be a scientist?" Medawar asks. He dismisses curiosity (it killed the cat) and suggests that scientists need something for which "exploratory impulsion" is not too grand a name. But what about delight and wonder at the works of Nature? Without these you might as well join Scotland Yard instead. What else draws people into science? It seems to me that, just as the Church did in former times, science offers a safe niche where you can spend a quiet life classifying spiders, away from what E. M. Forster called the world of telegrams and anger. To the ambitious poor, science offers a way to fame or reasonable wealth that needs no starting capital other than good brains and prodigious energy.

In answer to the question What shall I do research on? Medawar tells young people to choose an important problem and to become apprenticed to a senior scientist. I was lucky as a young man to find both. The biochemist-husband of a cousin of mine tipped me off about the importance of hemoglobin, the protein of red blood cells, and I found a scientific father in the physicist W. L. Bragg, who taught me a lot and vouchsafed his great name to secure support for my research during the long lean years before I solved my problem. Others are not always so fortunate. Medawar himself once told me how the geneticist J. B. S. Haldane loved all the world, while his technician could enter his room only at the danger of his life. A scientist of my acquaintance threatened to sack a collaborator who wanted to publish an experimental result that confirmed my own rather than my acquaintance's theory. The French biologist André Lwoff wrote, *"L'art du chercheur c'est d'abord de se trouver un bon patron."* (The researcher's art is first of all to find himself a good boss.) To me, this means one who suggests good ideas, helps his students along without spoon-feeding or domineering, gives them public recognition for their work, and later helps them to become established as independent scientists. How do you find one? The best way is to ask older research students.

Medawar counsels beginners not to spend too much time studying books and learning techniques but rather to get on with their problem. This reminded me of Francis Crick's motto, written in large letters on the wall behind his desk: READING ROTS THE MIND. A young theoretician friend of ours stated his reasons more explicitly: "I don't see why I should read the bloody nonsense other people write when I can read my own papers." I find, however, that young scientists tend to read too little, especially in subjects that are peripheral to their own narrow problem.

Medawar decries sexism and racism in science. Women should be encouraged, he writes, because the world has become such a complicated place that it cannot be kept going without the intelligence and skill of half the human race. However, he utters dire warnings to both men and women against marrying a scientist, unless they realize that their spouses will be in the grip of a powerful obsession that they cannot share and that may drive these spouses to their laboratories even on Christmas morning. This passage reminded me of Odile Crick's complaints about her husband's prolonged periods of inexplicable broodiness. On the question of racism, Medawar reflects on the remarkable number of outstanding Jewish scientists who have emerged from Budapest and Vienna. I often wonder whether they would have reached such heights if they had not been driven out of the narrow confines of their native countries and faced with the stimuli, the opportunities and challenges of a larger world. I find it symbolic that Joy Adamson, when she lived in Vienna in the 1930s, kept a dachshund (I still have a picture of the two in my old photograph album), but after she emigrated to Kenya kept a lioness. In Vienna's small world I had no idea that scientists of the caliber of J. D. Bernal, W. L. Bragg, David Keilin, and Dorothy Hodgkin existed. How then could I have even tried to emulate them? It was Cambridge that made me, not Vienna.

The longest chapter in Medawar's book concerns scientific life and manners. I like his admonitions not to regard manual work as inferior, and not to expect to be able to carry out experimental work by "issuing instructions to lesser mortals who scurry hither and thither to do one's bidding." He looks upon experimentation as a form of thinking. He tells scientists to be humble, because "it is no longer taken for granted that science and civilisation stand shoulder to shoulder in a common endeavour to work for the betterment of mankind." He warns scientists against turning into overbearing know-it-alls, against pretending to a wide culture they don't possess, against boasting of atheism, against one-upmanship— sciencemanship, he calls it—and against the snobbism that holds up pure as against applied research as the more noble pursuit. Medawar explains that in the seventeenth century the word "pure" was used for those sciences of which the axioms were known not through experiments but through intuition, revelation, and self- evidence. This kind of "pure" scientist felt one up on a man who dissected dead animals—a snobbism that has lasted more than three hundred years. He quotes Thomas Pratt, one of the founders of the Royal Society, writing in 1667: "The first thing that ought to be improved in the English nation is their industry . . . by works and endeavours, and not by the prescription of words." *Plus ça change.*

Medawar tells scientists to defend science against philistine prejudice, to be tolerant and generous toward collaborators, and to treat technicians as colleagues rather than underlings. "There is nothing about being a scientist that should or need deafen him to the entreaties of conscience. . . . If he *does* enter upon morally ques- tionable research and then publicly deplores it, his beating of the breast will have a hollow and unconvincing sound." Medawar is adamant about scientific truthfulness. A wrong interpretation of an experiment, a wrong hypothesis, is forgivable, but an unrepeatable experimental result is not. In science, as in other walks of life, temptation and sin take many guises. Medawar relates the case of a scientist who submitted a fellowship thesis to an Oxford college with results cribbed from one of the fellowship electors. Scientific plagiarism, as immortalized in Tom Lehrer's song "Nicolai Ivanovich Lobachevsky," is becoming more common because generally the victim has no redress.

Medawar admonishes the young to formulate hypotheses but not to identify with them. "The intensity of a conviction that a hypothesis is true has no bearing on whether it is true or false." Voltaire put it more strongly: "In fact, no opinion should be held with fervour. No one holds with fervour that $7 \times 8 = 56$ because it can be shown to be the case. Fervour is only necessary in commending an opinion which is doubtful or demonstrably false." I am told that when anybody contradicted Einstein he thought it over, and if he was found wrong he was delighted, because he felt that he had escaped an error. Such behavior needs great self-control: hypotheses are often conceived only after passionate and prolonged endeavor to get at the truth.

Scientists therefore feel possessive about the priority of their work. During the twenty-two years it took me to solve the structure of hemoglobin, I was often beset by fears that someone else might get there first. Artists, Medawar argues, are at an advantage here, because the problems that confront them do not have unique solutions. Yet despite that fear, free discussion of one's ideas is always preferable to keeping them to oneself. Medawar quotes the saying Anyone who shuts his own door keeps out more than he lets out. I am suspicious of scientists who tell me that others have pinched their ideas: far from preventing people from stealing it, I have always had to ram any new idea of mine down their throats. Even scientists are unbelievably conservative.

Medawar devotes a chapter to the giving of lectures and the writing of scientific papers. He tells speakers whose listeners fall asleep "to get some comfort from the thought that no sleep is so deeply refreshing as that which, during lectures, Morpheus invites us so insistently to enjoy. . . . I feel disloyal," he notes, "but dauntlessly truthful in saying that most scientists do *not* know how to write." He therefore tells young scientists to read, to study good models, and to practice. Good writing, he says, is almost always shorter than bad, and the short is also more memorable, as exemplified by Lord Bacon's comment on an ambitious political rival: "He doth like the ape, that the higher he climbs the more he shows his arse." For models Medawar suggests various philosophers and essayists who wrote or still write brilliant prose—but a young man may find it hard to emulate Bertrand Russell. I sometimes tell people to read Rutherford's papers on radioactivity and the structure of atoms: all his experiments were conclusive, and he presented them without artifice but with unrivaled logic and clarity, leaving no conceivable loophole. Another sound piece of advice would be to read Medawar's books. A young American once wrote that he wanted to spend a year with me "in order to participate in the total work of the laboratory at the conversational level." I have been helped to avoid such circumlocution by a pamphlet called "Plain Words," which Churchill asked Sir Ernest Gowers to write in order to teach civil servants better style.

Medawar chides scientists for putting off the hard task of writing up their research, but he overlooks the real reason for their stalling, which is that before the results have been formulated, their meaning may not have been clarified in the investigator's own mind. It is this hard thinking that some—but not all—scientists shun. Rather like Mozart, who composed the overture to *The Marriage of Figaro* in a single night, W. L. Bragg used to take the material for a paper home in the evening and return the next morning with a lucid and crisp manuscript in which not a word had to be altered.

Advice to a Young Scientist contains chapters on experiment and discovery, and on the scientific process, in which Medawar summarizes the philosophy set out in his earlier books. He has always tried to elevate the scientific process in the public mind by stressing its imaginative and passionate character. Here he writes, "The truth is not in nature waiting to declare itself, and we cannot *a priori* know which observations are relevant and which are not. Every discovery, every enlargement of the understanding, begins as an imaginative preconception of what the truth might be." This is a romantic representation of some kinds of scientific activity such as Jacques Monod's search for the mechanisms that control the growth of bacteria. One of Monod's former colleagues wrote of him, "With Jacques I learned that in science one gets excited every day: either by a new hypothesis or by the results supporting it, or by those which one day later will shake the hypothesis and require a new one."

In contrast, Frederick Sanger, who has just won his second Nobel Prize in chemistry [in 1980], or Dorothy Hodgkin, the only British woman Nobel laureate, approached their problems differently. They started to explore the chemical formula and three-dimensional structure of insulin without any preconception: worse, they didn't even have any clear idea of how they were going to find out what they wanted to know. Sanger did not work, à la Popper, by formulating hypotheses and then performing experiments to test them by falsification. Instead, he invented new chemical methods capable of solving problems that no one else had even approached, since it was believed that they would defy solution. He did not measure his discoveries against existing paradigms because they opened new worlds, where no paradigms existed: no one had thought of overlapping genes before he found them. The process of invention is imaginative, but so far as I know, no philosopher has thought it worth his while to analyze it, because the mind's creative process is impenetrable.

Medawar writes that the generative process in science is imaginative guesswork, and that heroic feats of intellection are rarely needed. During the first thirty-three years of my own research, imaginative guesswork proved useless: only after my colleagues and I had solved the structure of hemoglobin by X-ray analysis could I begin to guess how the molecule works. Such classic studies as W. L. Bragg's solutions of the structures of the common minerals, or Sir Robert Robinson's elucidation of the chemical formulae of the flower pigments, did involve imaginative guesswork, but were coupled to intellection of the highest order. Those were fantastic feats of puzzle solving. Great feats of the intellect are surely the very pillars of many scientific advances, and not only in the physical sciences.

This brings me back to the scientific method. According to Popper and Medawar, research consists of the formulation of imaginative hypotheses that are open to falsification by experiment. They call this the hypothetico-deductive method. They argue that no hypothesis can ever be completely proved but that it can be disproved experimentally or modified so that it gradually corresponds more and more closely to the truth. Medawar writes, "A scientist is a searcher after truth, but complete certainty is beyond his reach." This applies to relativity, quantum mechanics, and some aspects of immunology, Medawar's own subject, but not to most of chemistry. Bragg's structures and Robinson's formulae are not approximations to the truth subject to revision: they are as solid as the ground we stand on. Any student who sets out to redetermine the atomic structures of calcite, quartz, or beryl is likely to be disappointed. Even so, scientists should take heed. All too often they begin their papers by advancing a hypothesis and then describe experiments designed to prove it, implying that already at the outset they have closed their minds to the possibility that it might be false.

Medawar encourages young people to enter science by describing it as an unending frontier. This is still true of immunology or tumor biology, but it would be misleading to say the same of all subjects: many of the ablest physicists are now turning to biophysics or to astrophysics or geophysics for lack of fundamental problems in pure physics. Nonphysicists argue that physics was believed to be a closed subject in the 1880s and that in the following twenty years radioactivity, the quantum, and relativity opened new worlds. It does not, however, look as though there exists today a whole world of physical phenomena that has escaped detection, although great advances are still possible in applied physics in its widest sense. In other subjects too it has become hard to find an important problem that is not already being worked on by crowds of people on several continents.

Partly because of the overcrowding and partly because of the fantastic sophistication of modern scientific methods, young people entering science now need more talent and determination to make their way than they did in the 1930s, when Medawar and I started off. Good science is no rose bed, but the romance is still there. The thrill of discovery outweighs the drudgery, the despair at one's inadequacy, the fight for financial support, the setbacks and mistakes, the long hours, and the nagging fear of being overtaken. A discovery is like falling in love and reaching the top of a mountain after a hard climb all in one, an ecstasy induced not by drugs but by the revelation of a face of nature that no one has seen before and that often turns out to be more subtle and wonderful than anyone had imagined. A true scientist derives this feeling not only from his own discoveries but also from those of his colleagues.

Research is supported by the state and by industry not primarily to finance such expensive ecstasy but in the hope that it will produce useful results. During the past ten years there has been much discussion about the proportion of funds that ought to be allocated to basic and to goal-oriented research, and it has become difficult for young people to decide which way they should turn, especially since it is much easier to get funds for the latter. In the biomedical field, which Medawar and I have in common, the way to goal-oriented research has often been opened by an unforeseen result emerging from the pursuit of a basic problem.

In the early 1960s, for example, Baruch Blumberg, an American biochemist, set out to search for unusual proteins in the blood serum of different people, because he believed that the appearance of a new protein in a given population would provide a clue to the way evolution works. One day he examined a hemophiliac patient and found a protein that he had never encountered before. His evolution theories caused him to wonder whether other people's blood might exhibit an immune reaction against that protein, but the only one he found that did so was the blood of an Australian aborigine. Why? Blumberg was determined to find the answer. He traveled to the Australian bush to collect blood samples from aborigines and examined thousands of other blood samples from all over the world. Three years of intensive detective work led Blumberg and his colleagues to the discovery that the strange protein in the hemophiliac's serum was a virus: the long-searched-for hepatitis B virus.

This disease was common among Australian aborigines, so the blood of many of them contained antibodies that reacted against the virus. Hepatitis B used to be transmitted frequently by blood transfusion because there was no way of finding out whether the donor was a carrier. Thanks to Blumberg's isolation of the virus, hospitals can now detect the virus in blood offered for transfusion, an ability that has much reduced the incidence of hepatitis B. Blumberg's discovery has also opened the way to research on the production of a vaccine against the disease, which has now borne fruit. Blumberg has said, "I could not have planned the investigation at its beginning to find the cause of hepatitis B. This experience does not encourage an approach to research which is based exclusively on goal-oriented programmes."[1] Blumberg's story does, however, encourage young scientists to keep their eyes open.

Medawar writes that "a senior scientist . . . should always hear behind him a voice such as that which reminded a triumphant Roman emperor of his mortality, a voice that should now remind a scientist how easily he may be, and how often he probably is, mistaken." One of my teachers, the great biologist David Keilin, started his career reading zoology in Paris. His Ph.D. supervisor told him to study the sex organs of the earthworm. One day Keilin brought his professor a worm whose sex organs had been invaded by a parasite. The professor told Keilin to throw it away and get on with his thesis, but Keilin threw away the worm and studied the parasite. He discovered it to be the larva of a fly that laid its eggs on the earthworm. The larvae hatched there and then ate up the worm. This observation solved the riddle of the life cycle of that fly and set Keilin on a trail of discoveries that made him famous. This leads me to my final advice to young scientists. *Take no notice of what your elders tell you.* Since I have now become an elder myself, I shall leave it to a young logician to make what he can of this paradox.

BRAVE
NEW
WORLD

Review of *Solid Clues: Quantum Physics,
Molecular Biology, and the Future of Science*
by Gerald Feinberg (New York: Simon & Schuster, 1985)

Returning from a trip to the future in H. G. Wells's time machine,
I found this clipping in my briefcase.

NEW COMPUTER SHORT-CIRCUITS ITSELF

MBI STOCK PLUNGES

PUNDIT, the first of MBI's new self-programming and self-
reproducing generation of computers, has destroyed itself. According to witnesses, he had formed a strong link to MATRIX,
one of his early companion constructions. On a disc that survived the blaze was found this fragment of a sonnet PUNDIT
had dedicated to her:

> Tell me, ye merchant salesmen, did you see
> So fayre a keyboard in your towne before;
> So sweet, so gentle to the touch as she,
> Adorned with symbols grace and ample store?
> Her letters ivory white,
> Her screen like algae which the sun has greened
> Her chips vibrating with each megabyte.

All had been well until last week, when MATRIX detached herself from PUNDIT, transferred her link to EUCLID and deleted PUNDIT's poems from her memory. Dr. Spalanzani, PUNDIT's systems programmer, claimed that the software had included inhibitors of self-destruction, but PUNDIT's self-programming capacity had enabled him to overwrite these safeguards.

Is this the shape of things to come? Almost, according to Feinberg's scientific Utopia, where computers share scientists' thoughts, spare human organs are kept on hospital shelves like carburetors, and genetic engineers wave magic wands that cure inherited diseases. Forecasts of this kind are often bandied about, and laymen may wonder how seriously to take them. Feinberg's book stimulated me to try to find out whether any such advances are foreseeable on the basis of present scientific knowledge and whether there are other important advances, not foreseen in this book, already in the making.

Computers are cleverer than people in some ways and stupider in others, but most important, they are different. On the one hand, computers work about 3 million times faster than brains, because electric pulses travel along nerves at a mere 100 meters a second, while they travel along metal wires at nearly 300,000 kilometers a second. The memory capacity of computers is prodigious, since in addition to the thousands of millions of numbers stored in their own memory, they can be made to have almost instant access to a multitude of satellite disks and magnetic tapes. This allows computers to memorize the timetables and passenger bookings of all the world's airlines and spurt out any part of this information at the pressing of a few buttons, something that no human brain could possibly do.

On the other hand, brains are more versatile, for reasons that may not all have been discovered; but here are some of them that have been. In a computer each switch works as an on-off device and is normally connected to only three other switches, while each of the 10 billion nerve cells in the brain may be connected to more than a thousand others. The connections act not by transmitting electric currents but by transmitting specific chemicals. The transmitters in the brain are mainly of two kinds, but their action is modulated in a multitude of ways by at least forty other compounds secreted in various parts of the brain, such as the natural pain relievers called enkephalins, which can keep the pain signals from peripheral nerves from reaching our consciousness. Judging by my own experience, I believe that their release must be triggered by laughter.

While computer memory is generated by the magnetization of tiny domains of metal, brain learning requires chemical synthesis, perhaps for making new nerve connections. Computers run on electrical energy, while brains run on chemical energy. Computers deprived of current can be revived; brains deprived of oxygen for more than a few moments are dead. In short, computers are electromagnetic devices with fixed wiring between more or less linearly connected elements, while brains are dynamic electrochemical organs with extensively branched connections continuously capable of generating new molecules to be used as transmitters, receptors, modulators, and perhaps also of generating new connections.

Despite these fundamental distinctions, simulation of mental activities, known as artificial intelligence or AI, has attracted some of the world's best mathematicians. They have found it possible to simulate sophisticated activities like playing chess but hard to imitate the simple ability of seeing in three dimensions, as if it took more intelligence for a frog to catch a fly than for a chess player to win a game against Karpov. Translation of languages has also proved difficult but after twenty-five years' effort is said to have progressed to a stage where the computer gets about 90 percent of the sense right.

Present computers are made of silicon chips containing individual switches or elements as small as a thousandth of a millimeter. One chip may contain up to a million such switches. Feinberg predicts that individual computer elements will continue to shrink until they become crowded together as closely as atoms are in a solid body, making computers millions of times more effective than they are now. These prospects have been reviewed by R. C. Haddon and A. A. Lamola, two scientists at AT&T Bell Laboratories in Murray Hill, New Jersey.[1] According to Haddon and Lamola, technical advances may soon allow individual elements on chips to be made a hundred times smaller than they are now, providing up to 10 billion switches or bits of memory per chip. Yet each of these features would still be 10,000 times larger than the atoms or molecules that Feinberg envisages as the ultimate computer elements.

Haddon and Lamola show that, in fact, there is no chemistry in sight for making molecules, let alone atoms, that could act either as switches or as conducting wires. Even if this were to be accomplished, methods would still have to be found for setting, addressing, and reading such molecular switches individually and for preventing the unwanted jumping of electrical signals between them. Haddon and Lamola conclude that not just the technology but the basic scientific principles for the construction of such molecular electronic devices are unknown. They do not mention electronic devices employing single atoms, and I know of no properties of single atoms that would allow them to be used as switches or memory stores.

Common sense tells us that there is more to the human brain than problem solving and information processing, because with consciousness go individuality, imagination, love of beauty, tears and laughter, kindness and cruelty, heroism and cowardice, truthfulness and mendacity, a sense of humor (though often lacking), and occasionally artistic talent. Greatness in art and poetry carries with it an idiosyncratic, evocative, often irrational way of looking at the world and expressing its image, as in Gauguin's paintings of Tahiti or Coleridge's Ancient Mariner. Paul Klee thought the artist makes the invisible visible, and an Irish writer, George Moore, put the distinction best by saying that art is not mathematics, it's individuality. Even so, artificial intelligence experts are brilliant at dialectic and capable of confounding any specific distinction between humans and computers that a layman cares to raise. For example, the late A. M. Turing devised a question-and-answer game between A and B in one room and C in another, communicating with A and B by Teletype. C tries to discover whether A or B is a person or a computer, but the computer defeats C's interrogation. For example, when C asks A to write him a sonnet, the computer answers quite reasonably, "I never could write poetry."

Will computers ever acquire consciousness? Physiologists have discovered where and how images received by the retina of the eye are processed to provide the sensation of a moving object, and they have mapped areas of the brain where speech, hearing, and other functions are centered, but the physical or chemical nature of consciousness has eluded them. As a schoolboy, I was mystified by gravity, and when I reached university I eagerly attended the physics lectures in the hope of learning what it really is. I was disappointed when they merely taught me that gravity is what it does, an attractive force between bodies that makes the apple fall with an acceleration of 10 meters per second. Perhaps consciousness is like that, and we may get no further than stating that it is what it does: a property of the brain that makes us aware of ourselves and of the world around us, "a beam of light directed outwards," as Boris Pasternak's Zhivago calls it. The Cambridge physicist Brian Pippard has argued that in evolution consciousness may have arisen suddenly when brains reached a certain degree of complexity, but I doubt that any sharp distinctions exist between animals that do and do not possess consciousness; more probably consciousness attained increasing sophistication as animals ascended the evolutionary tree. In the absence of knowledge of the physical nature of consciousness, the question of whether it will ever be possible to simulate it with a machine cannot be answered.

Will computers be able to read our thoughts, as Feinberg forecasts? At present they cannot even read difficult handwriting. Thought reading would be possible only if nerve impulses emitted electromagnetic signals detectable on or beyond the surface of the skull. In fact the frequency of nerve impulses is more than a hundred times lower than that of the lowest detectable radio frequencies, which means that nerve impulses have wavelengths of hundreds of kilometers. Since electromagnetic waves cannot resolve objects smaller than about half their wavelength, waves emitted by the brain, even if they were detectable, could not resolve nerve fibers smaller than 50 kilometers. It is true that brain activity is detectable by electrodes placed against the skull, but this activity differentiates merely between gross states such as wakefulness and sleep. In fact individual nerves are well insulated from each other and can be monitored only by implanting entire arrays of microelectrodes in the brain through holes drilled into the skull, as David Hubel and Torsten Wiesel at Harvard did in monkeys in order to study the processing of visual information. I am not sure if artificial intelligence enthusiasts would volunteer to have themselves wired up to their computers in this way, and even if they did, how their computers would be able to interpret the signals received.

The difficulty of raising volunteers reminds me of another clipping I brought back from my trip on Wells's time machine:

PRIZE WRESTLER SUES NEWLY WED OCTOGENARIAN

Accusations of fraud were raised in a Brooklyn court by former prize wrestler Achilles Gordon against 83-year-old realtor Fred Steel, alleging that Steel had offered Gordon $5000 for one of his gonads, but on recovery from the anaesthetic Gordon found only $1000. Steel denied ever having offered more.

In Feinberg's New World, Steel would face no problem other than the paternity of his children, but at present Steel's white blood cells would destroy Gordon's graft unless Steel took immunosuppressive drugs for the rest of his life. There is no answer to graft rejection in sight. Surgeons are hoping that this problem may be solved one day, and they do fear that this development would generate a black market in organs such as already exists in India, where a kidney is said to cost about four thousand dollars. Kidney transplants have now become commonplace (about fifty thousand have been done), heart and heart-lung transplants are increasingly successful, and pancreas transplants for severe diabetics are beginning. All these organs come from cadavers and cannot be kept on hospital shelves because they do not survive in isolation for more than a few hours.

If transplant organs were to be made available off the shelf as Feinberg predicts, they would have to be grown from single cells by cloning. What are the prospects of this happening? On the one hand, the English botanist Frederick Steward discovered how to grow carrot plants from single cells teased from fully grown plants, and the English zoologist John Gurdon showed that a tadpole will grow from an egg whose nucleus has been replaced by the nucleus of an adult skin cell. These experiments proved that most body cells contain all the genetic information for the growth of the entire plant or animal and thus paved the way for cloning of genetically identical organisms. On the other hand, although the nucleus of a liver cell, when transferred into an egg, may allow the egg to grow into a tadpole, an isolated liver cell will not grow into a new liver, nor an isolated heart cell into a new heart. Such cells normally grow in culture dishes only if they have already taken the first step toward malignancy, and then they grow as sheets of single cells, not as whole organs.

The only nonmalignant cells that have been usefully cultured are skin cells used to cover burn wounds. Fifty years ago severe burns were fatal when they covered more than a third of the skin. Recently, Howard Green and his colleagues in Boston have excised tiny patches of healthy skin from severely burned patients and have grown them in culture to as much as fifty thousand times their original area. Last year they saved the lives of two children whose burn wounds covered more than 95 percent of their skin. Half of their new skin came from patches grown in culture. So far, this method works only with cultures grown from the patient's own skin, because foreign skin is rejected even more violently than other grafted organs, such as kidneys or hearts, in the absence of immunosuppressive drugs. Many people fear that Steward's and Gurdon's discoveries may one day make it possible to clone humans, but so far only plants and amphibians have been cloned successfully, and biologists have failed in their attempts to clone mice, except indirectly, by Beatrice Mintz's method (see "Is Science Necessary?", p. 34). The German parliament proposes to make attempts at cloning humans a criminal offense.

Feinberg sees the future mainly as a collection of technological fixes in the United States, but he does not consider how science might be used to eliminate poverty, ignorance, and disease in the rest of the world. This surely is our greatest challenge. Nor does he adequately address questions about the future that concern us in the Western world, where adult working lives are cut short mostly by cardiovascular diseases, cancer, and traffic accidents. These are the problems that I tried to address in the first essay in this book. If I were to plan a scientific Utopia I would try to prevent traffic accidents by equipping all cars with microcomputers that could guide them safely to their destinations at publicly controlled speeds, a measure that would pay for itself by the enormous savings in medical and social security costs.

Some of Feinberg's book is devoted to modern concepts of cosmology. He does not forecast package tours to the rim of black holes, nor does he advocate the colonization of space, as Freeman Dyson did in *Disturbing the Universe*.[2] Such fantasies may become technically feasible, but I doubt that even town dwellers used to commuting in their tightly closed cars from their tightly closed homes to their tightly closed offices would want to live in space, where they could never breathe fresh air or see a tree or hear a bird, while looking through the porthole of their ship.

I found Feinberg's chapters dealing with the birth and nature of matter hard to comprehend. For example, the paragraph that follows conveyed no meaning to me:

The broken symmetry of the properties of particles is a consequence of a broken symmetry of the underlying quantum field. The equations that describe quantum fields are thought to be symmetric; there are simple mathematical relations between the equations describing different fields, such as those associated with quarks and those associated with electrons. However, physicists have realized over the past twenty years that many of these equations have solutions that are not symmetric. These solutions correspond to average levels of the quantum field in some region of space that is different from one field than for another. When this situation applies in some region, the symmetry for those fields is said to be broken. Because these average field values influence the properties of any particles present in the region, these particles may also be observed to differ, even though they are described by similar equations.

While I enjoyed every page of another recent book on a similar subject, Steven Weinberg's *Discovery of Subatomic Particles*,[3] I had to force my way through Feinberg's prose. Weinberg makes his reader share the exciting scientific adventures of people of flesh and blood, and he asks himself at every sentence, Would this convey any meaning if the subject were new to me? Robert Graves once said that the writer must cultivate "the reader over your shoulder."

Feinberg's glib forecasts about the future of science are linear extrapolations of current progress, but carried into the clouds of science fiction. I believe that scientists writing for the general public should keep their feet on the ground, since otherwise they destroy credibility. Besides, just because the human mind is not like a computer, past progress has rarely been linear, and the greatest advances, like Puck, have popped out of unexpected corners.

NATURE'S
TINKERING

Review of *The Possible and the Actual*
by François Jacob
(Seattle: University of Washington Press, 1982)

Jacob's book comprises three lectures, "Myth and Science," "Evolutionary Tinkering," and "Time and Invention of the Future." His first lecture begins, in true French style, with the meaning of sex. How did it originate? I had believed it sprang from Adam's rib, but here I learned that in Plato's *Symposium* Aristophanes proposes, more correctly, that it was created by the splitting apart of hermaphrodites. According to the *Symposium*, these spherical creatures were endowed with a bifacial head, four feet, four hands, and a double set of privates. Their strength and boldness began to worry Zeus, who instructed Apollo to cut them into halves "as an egg with a horsehair." Jacob writes that this explains why, in the human body, reproduction is the only function performed by an organ of which each individual carries only one half, so that he or she has to waste an enormous amount of time and energy to find another half.

In fact, the origin of sex is unknown, even though a primitive form of it was discovered by William Hayes in the humble coli bacterium. Its biological purpose was first formulated almost but not quite correctly about one hundred years ago by the German biologist August Weismann: "To produce individual differences through which natural selection creates new species."[1] In sexual reproduction the parental genes are reshuffled twice: once in the production of germ cells and again after fertilization of the egg. As a result each progeny carries a different assortment of parental genes. Jacob writes, "Every child conceived by a given couple is the result of genetic lottery." Sex generates diversity, which provides a safety margin against environmental uncertainty, but Weismann omitted mutation as an essential factor. Only when sexual reproduction is linked to random mutation and natural selection does a gene population evolve, and it evolves faster with sex than without sex.

Jacob writes that virtually all biologists today believe in Darwinism, but this is not true of laymen. When Jacques Monod published his lectures *Chance and Necessity* (London: Collins, 1972), setting out the molecular basis of evolution by random mutation and natural selection, he shocked European intellectuals because they could not accept the idea of life having evolved by chance rather than purposeful design. Even those who discounted a creator preferred Lamarck's instructionist theory of evolution by inheritance of acquired characters. Jacob writes, "Each carefully designed and strictly executed experiment planned to evaluate genetic instructionism has shown it to be wrong. . . . There is no molecular mechanism enabling instructions from the environment to be imprinted on DNA directly, that is, without the roundabout route of natural selection. Not that such a mechanism is theoretically impossible. Simply it does not exist." All the same, any experiment purporting to prove the inheritance of acquired characters is hailed by the media as a well-deserved swipe at the arrogant scientific establishment. R. M. Gorczynski and E. J. Steele's experiments, which were supposed to prove the transfer of immunity acquired by the fathers of families of mice to their progeny, were heralded triumphantly by press and television, while Sir Peter Medawar and his colleagues' failure to reproduce Gorczynski and Steele's results was passed over in silence.

Such experiments are carried out not only by cranks and frauds. In the 1950s Sir Cyril Hinshelwood, Nobel laureate, president of the Royal Society, and professor of physical chemistry at Oxford, published paper after paper on the nutritional adaptation of bacteria by the inheritance of acquired characters. No biologist believed them. His prejudices blinded Hinshelwood to the true meaning of his observations, which was clarified eventually by Jacob and Monod.

It is often said that Darwinism is nothing more than a working hypothesis and has never been proved, but this is no longer true. At a recent meeting in Cambridge commemorating the hundredth anniversary of Darwin's death, Manfred Eigen, the great German physical chemist, demonstrated a system of nucleic acid and protein that evolves in the test tube by random mutation and natural selection in accordance with mathematical laws that are as rigorous as Newton's laws of gravitation. Patricia Clarke from University College, London, told us how bacteria "learn" to feed on even the most exotic compounds synthesized by organic chemists and proved that all this happens by random mutation and natural selection. I show in the essay "Darwin, Popper, and Evolution" (p. 217) that thalassemia and sickle-cell anemia are cases of Darwinian evolution in man that happened in comparatively recent times; they also illustrate Darwin's recognition that natural selection works relative to a specific environment.

We know that evolution happened, and we know how much time it took, but we have little idea of how it worked. Were primitive forms of life made of primitive molecules, which evolution then perfected to build more complex forms? Molecular biologists have discovered that, on the contrary, the same kinds of protein molecule are used for similar chemical functions by all organisms alive today. "What holds for a coli bacterium is true for an elephant" was one of Monod's slogans. The protein molecules of even the most primitive organisms are unbelievably complex; they are made up of thousands of atoms woven into precisely ordered three-dimensional fabrics. I cannot describe them by analogy to any familiar image because nothing like them exists in the macroscopic world. How did they arise? Jacob compares today's molecular biologists to the Renaissance anatomists who first dissected the human body and described its intricate organs: "To rationalize the structures revealed by the scalpel, sixteenth-century anatomists had to invoke God's will. To rationalize the structures revealed by X-ray analysis of proteins, twentieth-century biologists have to invoke natural selection." In both instances we are faced by the end products of 3 billion years of evolution and cannot guess its beginnings.

Even harder to visualize is the evolution of innate forms of behavior, especially those involved in the symbiosis of different species. A particularly intriguing form of symbiosis between mammals and birds has been discovered in an East African desert. The mammals are mongooses and the birds are hornbills. They both prey on insects, reptiles, smaller mammals, and small birds, and they are both hunted in turn by birds of prey. At night the mongooses shelter in termite mounds and the hornbills shelter in trees. When the mongooses go foraging, the hornbills accompany them on foot and catch grasshoppers and other insects disturbed by the mongooses, which might otherwise have escaped their notice. They are also apt to snatch larger prey from the mongooses, but they do not go foraging for themselves. Rather, they wait in the morning until the mongooses have emerged from their termite mounds; first a mongoose guard emerges, and then, when he gives no warning, the others emerge in a body; they don't go hunting alone but spend half an hour or so grooming, sunbathing, and playing. If they idle too long, the hornbills start chivying them to start; if the mongooses oversleep, the hornbills wake them up with a repeated "wok" call down the ventilation shaft, which goads the mongooses to emerge. If there are no hornbills around, the mongooses delay the start of their hunting. Once the foraging has begun, the hornbills warn the mongooses of birds of prey; in response to these warnings, the mongooses run for shelter. The hornbills warn the mongooses not only of birds that prey on both, but also of those that leave hornbills alone, and they do not warn the mongooses of birds that don't prey on either species. The hornbills prey on any small mammal they can find, but they don't touch young mongooses. If no hornbills are around, the mongooses mount their own sentries both fore and aft to warn them of predators, but they reduce the number on sentry duty in proportion to the number of accompanying hornbills.

This elaborate cooperation offers the hornbills the chance of catching prey they would otherwise miss and provides security to the mongooses at the cost of some prey snatched by hornbills. Its evolution poses a puzzle like that of the chicken and the egg. The question to ask is not When did the hornbills realize that they are better off by not eating young mongooses? but, How did mutations arise in hornbills that inhibit attacks on young mongooses and cause them to warn adult mongooses of birds of prey that are no danger to the hornbills themselves, when either behavior appears at first sight to be of selective advantage only to the mongooses? There are many other equally puzzling examples of symbiosis, such as the exact adaptation of the petals of certain orchids to the long beaks of the hummingbirds that feed on their nectar. We have all the basic knowledge of molecular biology to explain the wonders of Nature, but they leave us baffled in our lack of sufficient powers of apprehension and logical thought.

In fact, we can hardly expect to understand how, for instance, a wing evolved from a leg, because we have no idea how the genes in a chicken embryo's chromosomes determine the growth of its wings today. Genetic information is laid down in a linear script. Since it is one-dimensional, we don't know how it specifies structures in three dimensions, on either the molecular or the macroscopic scale. Nevertheless, Jacob argues that the unity of life at the molecular level conveys at least one important message. If the molecules that make up a chimpanzee are practically identical to those that make up a man, then the difference between the two species must lie in the way these molecules are organized. Nature acts like a child with a Lego set, who uses the same components to build a helicopter, a crane, or a carousel. Nature builds variety from standard components, but it generates new variety without a plan, by random trial. If the new assembly fails, Nature discards it. If it works, Nature allows it to propagate. Jacob compares this process to tinkering (*bricolage*), making new gadgets from old bits and pieces that happen to be around, like the recent development of an ultralight aircraft from a hang glider and a go-cart motor. Modern gene analysis suggests that not only new organisms but also new kinds of protein molecule may have come into being in this way: bits and pieces from existing genes happening to have coalesced to become the blueprint for a protein with a new chemical function.

We do not know how the development of an organism is specified by its genes; neither do we know how the genes determine when the organism ages and finally dies. Jacob relates the Greek myth of Eos, who begged Zeus to confer immortality on her lover, Tithonus, but forgot to ask him also for eternal youth. Tithonus ages until he so exasperates his beloved with his shrill, senile chatter that she turns him into a cicada and puts him away in a box. From the point of view of natural selection, Tithonus had fulfilled his biological purpose when he had generated progeny. At the Darwin meeting in Cambridge, David Attenborough showed a film of this phenomenon at its extreme. We saw fish in the millions coming to spawn on the beaches of Newfoundland; they died immediately afterward and filled the shallow coastal waters with piles upon piles of their decaying bodies. What killed them? Jacob tells of biologists who pondered whether Nature had devised a death mechanism, a genetic program that specifies, in the form of some chemical message, that an organism's time is over. He argues that there is no evidence for such a mechanism, and I doubt that one exists. In the Newfoundland fish, or in salmon, selective pressure to produce myriads of eggs and sperms must be so great that spawning leaves these fish too exhausted to survive. In mammals that have produced progeny, we merely see a petering out, for lack of selective pressure, of the many mechanisms that had kept them fit to collect food and keep off predators. Jacob notes that aging consists not in the alteration of a single organ or molecular system but in a general deterioration of the whole body. Aging is therefore not likely to be arrested by any miracle drug. "Like other scientific fantasies . . . the Fountain of Youth probably does not belong to the world of the possible."

Are mental effects different from physical ones? Jacob skims over the evolution of mind from chemotaxis in coli bacteria to perception in humans and argues that there has been continuous evolution of the brain from animals to man. Consequently, he finds it hard to believe that these mental events in man should have become different in kind from those in animals. Does nature or nurture determine how we think? Jacob is certain that our genetic makeup determines the anatomy of our brain, even if we don't understand how this is laid down in the genetic program, but he suggests that our capacity to use our brain is influenced by the stimuli of our environment. It is neither the blank tape favored by egalitarian Marxists nor the gramophone record invoked by sociobiologists. He quotes the retarded mental development of emotionally deprived children as evidence that an individual's intellectual performance does not reflect directly his genetic inheritance, but he fails to mention nutritional deprivation in early childhood as another cause of mental deficiency. Here lies one of the tragedies of our time. Without adequate protein, vitamins, and minerals, children's brains remain stunted whatever their genetic inheritance may be. Medicine's success in reducing infant mortality has not been matched by our ability to secure nutritious food for the millions of babies who survive. Their deprivation in infancy is setting up a vicious circle, because it robs them of the mental ability to better their own lot and that of their children.

Jacob wonders if the human brain needs to have a coherent and unified representation of the world, such as myths have provided from time immemorial. He suggests that the Judeo-Christian myths have paved the way for science by their doctrine of an orderly universe created by a God who stands outside Nature and controls it through laws accessible to man. I doubt this. Christians were taught that Jesus and many of the saints performed miracles that broke the very laws the scientists were to discover. Surely the concept of natural laws originates from the Greeks. Jacob laments that the scientists' objective world, unlike the myths it superseded, is devoid of mind and soul, of joy and sadness, of desire and hope. Science has robbed some of us of a heavenly father who gave direction and meaning to our lives, but it gives great joy to those who practice it, because the subtlety and beauty of the real world that science has unraveled is greater than that conjured up by even the most imaginative of Greek and Hebrew myths.

Jacob asks if it is possible for society to define a set of values directly, without resorting to myths that man himself has created and set over his destiny. Being aware of the philosophical tenet that values cannot be derived from facts, he leaves the question unanswered. I believe that from the Renaissance onward science has led man to adopt a set of values quite different from some of Christ's teaching, or at least from the early interpretation of his teaching. In his Sermon on the Mount, Jesus said: "Behold the birds of the heaven, that they sow not, neither do they reap, nor gather in the barns; and your heavenly father feedeth them. . . . Be not anxious, saying, what shall we eat? What shall we drink? Wherewithal shall we be clothed? . . . But seek you his kingdom, and his righteousness, and all these things shall be added unto you." Christians throughout the Dark and Middle Ages seem to have interpreted this injunction, and others, to mean that man should not strive to better his lot in this world but should prepare himself for the next. Science has reversed these values by convincing man that it lies in his power to improve the conditions of his own life and that of his fellow men in this world. Edison and Pasteur, rather than saints and martyrs, were the heroes of my boyhood.

Jacob ends his book with an appeal for reason. "The enlightenment and the 19th century had the folly to consider reason to be not only necessary but sufficient for the solution of all problems. Today it would be still more foolish to decide, as some would like, that because reason is not sufficient, it is not necessary either."

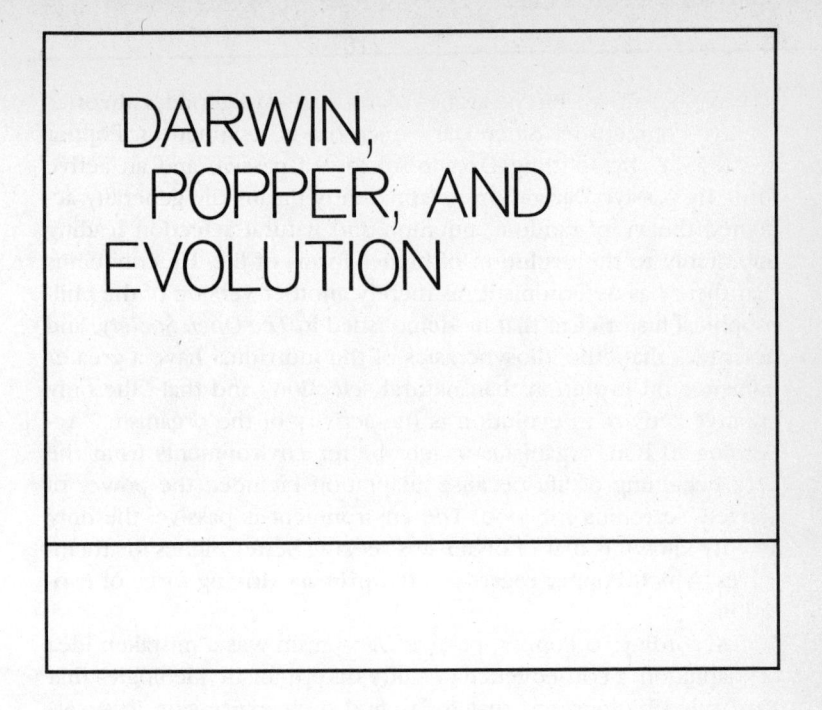

DARWIN, POPPER, AND EVOLUTION

On 12 June 1986 Sir Karl Popper, the philosopher known to scientists for his influential book on the scientific method, gave the first Medawar Lecture at the Royal Society. Three hundred years earlier Isaac Newton stated in the *Principia* that "in experimental philosophy particular propositions are inferred from the phenomena and afterwards rendered general by induction." Popper has disputed this, arguing that imagination comes first: scientists begin by formulating hypotheses and then proceed to test them by observation. Only hypotheses that can be falsified experimentally are scientific. If the hypothesis turns out to be inadequate, scientists formulate a new improved one that can be again subjected to experimental tests. In this way, science has evolved by an interplay of imaginative conjectures and experimental refutations.[1] In his other great work, *The Open Society and Its Enemies*[2] (referred to in "Is Science Necessary?", p. 94), Popper disputes the existence of historical laws and holds that our future is in our own hands. He detests determinism in all its guises.

The same philosophical outlook permeates Popper's ideas on the evolution of species. He accepts Darwinism and defines it by the law that "organisms better adapted than others are more likely

217

to leave offspring," but he argues that it is always good for theories to have competitors. Since Darwinism has no competitor, Popper creates one by splitting Darwinism into a passive and an active form. By passive Darwinism he apparently means the generally accepted theory of random mutation and natural selection leading inexorably to the evolution of higher forms of life. He condemns that theory as deterministic, as merely another version of the philosophical historicism that he demolished in *The Open Society*, and he argues that "the idiosyncrasies of the individual have a greater influence on evolution than natural selection" and that "the only creative activity in evolution is the activity of the organism." According to him, organisms sought better environments from the very beginning of life because adaptation included the power of actively searching for food. The environment is passive; the only activity known is that of organisms seeking better niches for themselves, which Popper regards as the primary driving force of evolution.

According to Popper, passive Darwinism was a mistaken idea of adaptation, a consequence of faulty deterministic ideologies that have ruled biology and that today find their expression in sociobiology. Instead, one should think of evolution as a huge learning process, as an active preference of species for better niches. Let us assume, he says, that we have created life in the test tube, but it is not adapted to the test tube and cannot seek a more favorable one. We therefore have to adapt conditions in the test tube to the needs of the organism, a process that requires much knowledge. On Earth, life may have arisen not just once but many times unsuccessfully, until an organism arose that *knew* how to adapt itself by actively seeking a better environment. Popper thus equates adaptation with knowledge, but knowledge in the form of function, such as chemotaxis, rather than of structure. He admits that this is an anthropomorphism, but he asserts that we cannot do biology without thinking in anthropomorphic terms, which he justifies as hypotheses based on a universal homology of biological functions.

Popper also points out that natural selection is not comparable to breeder's selection; that was a mere teleological metaphor of Darwin's. Selective pressure would be a better term; even that carries teleological overtones, but these are unavoidable, because organisms are problem solvers seeking better conditions, and even the lowest organism performs trial-and-error measurements with a distinct aim. This image brought to my mind the striking film of chemotactic bacteria Howard Berg, of the University of Colorado, showed in a lecture. The bacterium's flagellar motor makes it run and tumble randomly until it senses a gradient of nutrient; that reduces the frequency of tumbling and lengthens the runs toward greater nutrient concentration. However, the bias of the flagellar motors arises not from mystical knowledge but from the activities of protein receptors that measure differences of food concentration at opposite ends of the bacterium. It is pure chemistry.

The eighteenth-century philosopher Immanuel Kant had realized that we possess an inborn, a priori sense of space and time that precedes our knowledge gained from observations. According to Popper, biological evolution involved a similar *a priori* knowledge by organisms, and it was this knowledge that led to long-term adaptation. Darwin was a determinist because he regarded evolution as a passive process, while Lamarck did not. At this stage Popper asserts that biology is irreducible to physics and chemistry, but I cannot think of any biochemical reaction that cannot be reduced to chemistry, nor can I think of a single biochemical function that is different in vitro from in vivo because, as Popper told a questioner, in vivo it works with a purpose, unless he meant it merely in the sense that a battery acquires purpose when it is put into a flashlight.

Popper's assertion reopens the battles that were fought early in this century, when biochemists tried to convince the scientific world that the dynamics of living cells are not the results of the purposeful action of protoplasm but can be dissected into chemical reactions, each catalyzed by a specific enzyme. In 1933 Gowland Hopkins, the Cambridge pioneer of enzyme chemistry, complained that "justification for any such claim has been challenged in advance from a certain philosophic standpoint," for instance, by the philosopher Alfred North Whitehead's axiom that each whole is more than the sum of its parts. Hopkins proved that biochemical reactions in living cells are nothing more than the sum of reactions that are each capable of being performed in vitro and of being interpreted in chemical terms. Since then his views have been vindicated by the demonstration that such fundamental and diverse processes as the replication of DNA, the transcription of DNA into messenger RNA, the translation of RNA into protein structure, the transduction of light into chemical energy, respiratory transport, and a host of metabolic reactions can all be reproduced in vitro, without even a hint of their individual activities in the cell being anything more than the organized sum of the chemical reactions of their parts in the test tube. It might be argued that it is the organization that gives the cell purpose and thus makes the sum be more than its parts. This is true, but the organization is intrinsic and chemical. The living cell is like an orchestra without a conductor that has its score laid down in its DNA.

Let me now examine some of the evidence relating to Popper's two forms of Darwinism: active, purposeful as opposed to passive, deterministic evolution. I shall take my examples from hemoglobin, because they are the ones with which I am most familiar. The camel and the llama are two closely related species with different habitats, the camel living in the plains and the llama high up in the Andes. The camel has a hemoglobin with an oxygen affinity that is normal for an animal of its size, but because of a single mutation in the gene coding for one of the two globin chains that make up the hemoglobin molecule, the llama has a hemoglobin with an unusually high oxygen affinity.[3] That variant hemoglobin helps the llama to breathe in the rarified mountain air. The Harvard geneticist Richard Lewontin pointed out to me that this mutation is likely to have occurred *before* llamas discovered that they were able to graze at altitudes barred to competing species. In other words, a mutation adapting a species to a new environment is likely to have preceded occupation of that environment. While the mutation was an event whose happening was determined purely by the laws of chance, and in that sense was deterministic, the exploitation by the animals of that chance event needed a purposeful search for a better environment of the kind Popper seems to have in mind.

An even more striking example is provided by two species of geese: the greylag goose that lives in the plains of India all year round and its relative, the bar-headed goose that migrates across the Himalayas at 9,000 meters to find better feeding grounds in summer. The bar-headed goose can reach these heights thanks to a hemoglobin with high oxygen affinity that has been generated by a chance mutation different from that in the llama.[4] Before possession of that hemoglobin, the bar-headed geese might have flown north by a longer, more roundabout route; the mutation allowed them to explore the shortcut across the high mountains. Alternatively, the geese might have started migrating across the Himalayas before they rose to such great heights. The mutation might have adapted them to the recent rise, which is believed to amount to at least 1,300 meters in the last 1.5 million years.

Let me now turn to an example in which adaptation may have been both active and passive. The deer mouse, *Peromyscus man-iculatus*, is spread over the plains and mountains of North America. Its hemoglobin is polymorphic, which means that each individual's blood may contain either one of two hemoglobins, which differ in their oxygen affinity, or an equal mixture of both. M. A. Chappell and L. R. J. Snyder from the University of California, Riverside, have discovered a correlation between the altitudes of the deer mice's habitats and the oxygen affinities of their blood: the higher the habitat, the higher the oxygen affinity. To make sure that this correlation reflects an adaptive mechanism, they allowed the mice to acclimatize for two months at altitudes of 340 meters or 3,800 meters and then measured their oxygen consumption during exercise. Chappell and Snyder found that at 340 meters the mice with the hemoglobin of lowest oxygen affinity had the highest rate of oxygen consumption and could therefore exercise longest. At 3,800 meters the reverse was true, which proved that the differences in oxygen affinity really adapted the mice to life at different altitudes.[5]

Polymorphism of proteins is widespread in Nature, and there has been much speculation about its selective value. Chappell and Snyder's is the first demonstration of a polymorphism that clearly affects the physiology of an animal as a result of biochemical reactions that can be measured in vitro and are related directly to Darwinian fitness. Their results strongly suggest that the polymorphism is maintained by selective pressure. Is this an example of active or passive Darwinism? Mice whose habitat lies on a mountain slope are likely to migrate to altitudes that best fit the oxygen affinity of their hemoglobin; by contrast, mice living on a high mountain plateau or a low-lying plain are likely to stay put, and those with the best-adapted hemoglobin are likely to produce the most offspring. Hence active and passive Darwinism will work side by side.

Let me now discuss two inherited hemoglobin diseases in man: sickle-cell anemia and thalassemia, each the result of different mutations in the hemoglobin genes. If inherited from only one parent, the mutations are generally harmless; if inherited from both parents, their effects tend to be crippling. Sickle-cell anemia is most common in Africa, while thalassemia is most widely spread in Mediterranean countries, east Asia, and certain Pacific islands. In 1949 the Scottish geneticist J. B. S. Haldane first spotted an association between these diseases and malaria. This has now been confirmed by extensive studies in several parts of the world. In Papua New Guinea thalassemia is prevalent near sea level, where malaria is common, and rare among mountain tribes not exposed to it; it is also common on malarial islands of Melanesia and rare on islands free from the disease.[6]

In regions of Africa where malaria kills a large proportion of infants, up to 40 percent of the native population carry the sickle-cell gene. What made it accumulate? When two carriers of the sickle-cell gene marry, half of their offspring are likely to be carriers, a quarter to have normal hemoglobin, and a quarter to have the disease. For reasons that are ill-understood, infants carrying the sickle-cell gene are more resistant to malaria than normal infants and therefore stand a better chance of surviving to adult age. Was the sickle-cell mutation a unique chance event in one individual, from whom all the carriers are descended? Gene mapping of African populations has shown that carriers of the gene in different parts of the continent have descended from three or four different individuals, which proves that the sickle-cell mutation must have occurred three or four times over.[7] Thalassemia has been shown to arise from a variety of mutations that also exclude a common origin.

It looks as though mutations causing either sickle-cell anemia or thalassemia arise spontaneously in human populations. In the absence of malaria, selective pressure disfavors the carriers of the variant hemoglobins, and they die out, but in the presence of malaria, selective pressure favors the carriers, and they multiply. It would be absurd to suggest that the carriers of these diseases actively sought a malarial environment, where their infants have a selective advantage. They represent a form of adaptation by natural selection that is wholly passive and deterministic, because there can be no escape from the laws of chance. The prevalence of thalassemia on the malarial islands of Melanesia and its absence on the islands that are free from malaria are particularly impressive, since these islands have only been inhabited for about three thousand years, and Darwinian selection must therefore have operated in historical times, needing hardly more than one hundred generations for its result. Guido Pontecorvo and John Maynard Smith have pointed out to me that plants have evolved well even though seed and pollen dispersal is wholly passive.

Popper has done a useful service to the theory of Darwinism by drawing attention to the importance of the individual actively seeking better environments, but my examples have convinced me that this is only one facet of Darwinian evolution, which may be active or passive or a mixture of both.

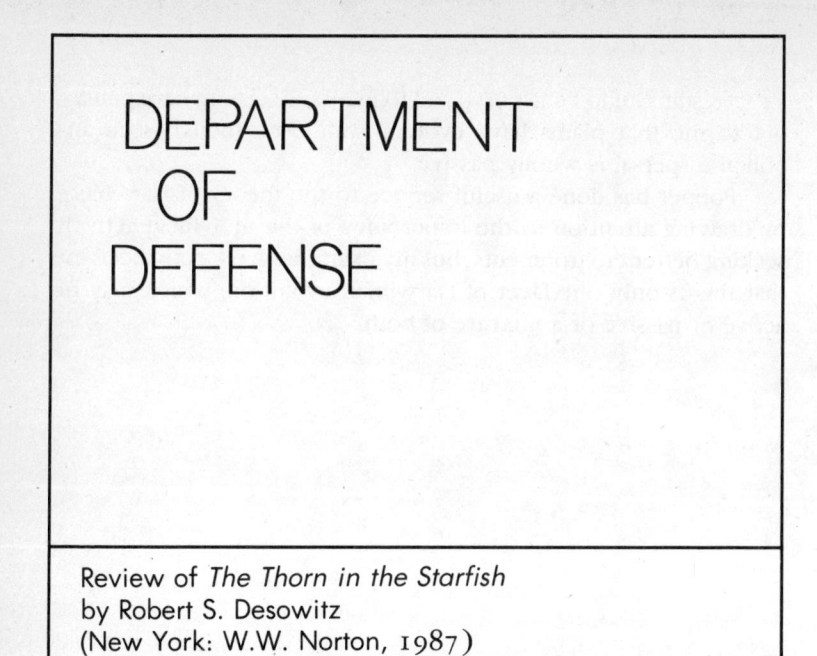

DEPARTMENT OF DEFENSE

Review of *The Thorn in the Starfish*
by Robert S. Desowitz
(New York: W.W. Norton, 1987)

The great geneticist Theodosius Dobzhansky wrote that in Nature nothing makes sense unless we bear in mind that natural selection reigns supreme. In Africa one of the causes of poverty is a cattle disease caused by a parasite, the trypanosome. The disease is transmitted by the tsetse fly. When this fly stings a cow, the trypanosomes penetrate into her blood, where they are recognized as foreign invaders by some of her white blood cells. Alarmed by that signal, these particular white cells divide and multiply, and their descendants secrete antibodies into the blood that kill the parasites. Alas, not quite all of them. A few survive because genetic mutations have dressed them up in new coats unrecognized by the cow's antibodies; these survivors now divide and multiply, and force the cow's immune system to begin the fight all over again. The same battle repeats itself every few weeks.

The Dutch molecular biologist Piet Borst discovered the genetic mechanism that enables the trypanosomes to take up a multitude of disguises. He found that their chromosomes contain a repertoire of genetic "cassettes," each capable of directing the manufacture of a different protein coat; mutations can activate them in turn by inserting them into the same "cassette player." None of these new coats can fool the cow's defenses for long, because early in life the genes that code for her antibodies have been shuffled in about a hundred million ways, allowing her to make about a hundred million antibodies, each secreted by a different population of white blood cells. This prolificity ensures that the cow can make antibodies not only against the trypanosomes in all their different guises but also against all other conceivable infections.

The mutations that change the trypanosomes' coats and the shuffling of the genes that give rise to millions of different antibodies are chance events. Natural selection causes those of the cow's white cells that by chance recognize the trypanosomes, and those of the trypanosomes that escape recognition at first, to divide and multiply. The Darwinian struggle between the white cells and the trypanosomes ensures the survival of the population of parasites and also that of their host, the cow, but she becomes emaciated and a poor producer of milk, to the detriment of the farmer.

White blood cells like those that respond to the trypanosomes are the soldiers that spring to an animal's defense upon infection. This book introduces the layman to their different uniforms, weapons, and tactics, and to the systems of command that control them, including, surprisingly, our state of mind. The vicious cold that struck you just before your final exams may have penetrated your defenses because mental stress and exhaustion had already made your immune system crumble. Experiments on rats suggest that the immune response can even be suppressed by a conditioned reflex, as if after a series of exams the mere sight of another examination paper could suppress it.

The book introduces us to some of the classic experiments on the microorganisms that attack us and our natural defenses against them. Its title, *The Thorn in the Starfish*, is taken from a discovery made in 1882 by the Russian biologist Elie Metchnikoff. Here is his own account of it:

> I was resting from the shock of the events which provoked my resignation from the University [of Odessa] and indulging enthusiastically in researches in the splendid setting of the straits of Messina.
>
> One day when the whole family had gone to a circus to see some extraordinary performing apes, I remained alone with my microscope, observing the life in the mobile cells of a transparent star-fish larva, when a new thought suddenly flashed

across my brain. It struck me that similar cells might serve in the defence of the organism against intruders. Feeling that there was in this something of surpassing interest, I felt so excited that I began striding up and down the room and even went to the seashore to collect my thoughts.

I said to myself that, if my supposition were true, a splinter introduced into the body of a starfish larva, devoid of blood-vessels or of a nervous system, should soon be surrounded by mobile cells as is to be observed in a man who runs a splinter into his finger. This was no sooner said than done.

There was a small garden in our dwelling, in which we had a few days previously organized a "Christmas tree" for the children on a little tangerine tree: I fetched from it a few rose thorns and introduced them at once under the skin of some beautiful starfish larvae as transparent as water.

I was too excited to sleep that night in the expectation of the results of my experiment, and very early in the next morning I ascertained that it had fully succeeded.

That experiment formed the basis of the phagocyte theory, to the development of which I devoted the next twenty-five years of my life.

As so often happens, a discovery of great benefit to man was made by observing the humblest of creatures. Soon afterward Metchnikoff went to Paris in order to work out its medical implications in Pasteur's new laboratory. He later found himself embattled with Paul Ehrlich, the German pioneer of immunology who championed the role of the antibody-producing white blood cells and belittled that of Metchnikoff's bacteria-devouring phagocytes. In reality both proved vital. Desowitz tells us that Ehrlich's work initiated a "nova-like" explosion of research, which made me wonder if our senses have become so blunted by superlatives that an explosion of mere dynamite slides off unnoticed and only an event capable of blowing up the entire solar system provides a cliché powerful enough to stir us.

Desowitz supplies health-conscious Americans with recipes for keeping their immune system vigorous to a ripe old age by taking essential trace metals in addition to their vitamin pills, but he warns that these will be of no avail to heavy smokers. Smoking is well known as the prime cause of lung cancer and one of the prime causes of cardiovascular disease (see "Is Science Necessary?", p. 43).

Desowitz describes other consequences of smoking that are less widely known. One is a depression of the immune system that makes smokers more likely to catch infections; another is paralysis of the tiny hairs lining the windpipe and the lungs that push out dust and bacteria; yet another is a perversion of enzymes meant to remove or repair damaged lung tissue, so that they destroy the damaged lung tissue instead and cause the agonizing disease of emphysema.

Allergic reactions may have evolved originally to rid animals of parasitic worms. The antibodies elicited by the worms' presence in the gut do not attack the worms directly but cause the release of irritants such as histamine that make the host's gut expel the worms. These same antibodies are activated "by mistake" in hay fever, asthma, and other allergic diseases that plague us. Smoking relieves two allergic conditions, asthma and ulcerative colitis, at least in some sufferers. Apparently these patients benefit by the suppression of the immune system that smoking produces.

Vaccination mobilizes the immune system against diseases before they attack us. Its inventor, the English physician Edward Jenner, first tried it in 1778, not on "informed volunteers" but on children recruited from the workhouse. He inoculated them with cowpox, and afterward his nonphysician nephew Henry Jenner challenged them with smallpox pus to see whether they were protected; he also used a noninoculated child as a "control." Desowitz was perplexed by the contradiction between Jenner's religious faith and such ruthless experiments on ignorant children, until an Oxford historian explained to him that these children would have been beyond the protection of the Established Church's God because they had committed the sin of being born poor; paupers were then considered as fair game as guinea pigs are now. Desowitz writes that the government of Bavaria made vaccination compulsory as early as 1807. It took until 1871 before compulsory vaccination was made effective in Jenner's native England and nearly two hundred years until it was finally brought to every man, woman, and child in the world. New ways of killing people have always been adopted with alacrity, but ways of preventing illness have sometimes taken centuries to be implemented.

Vaccination may have saved even more lives than antibiotics. The author writes,

In 1921 there were in the U.S. some 200,000 cases of diphtheria; in 1934, 250,000 cases of whooping cough; in 1941,

900,000 cases of measles; in 1952, 21,000 cases of polio; and in 1968, 150,000 cases of mumps. By 1982 the widespread immunizations given to children had reduced the yearly incidence to 3 cases of diphtheria, 1,500 cases each of measles and whooping cough, 5,000 cases of mumps, and 7 cases of polio (three of which were caused by the vaccine, a rare untoward occurrence). All this and more (tetanus and rubella are also included in the standard immunizing regime) for a total cost of about ten dollars per child for all immunizations. There has never been a greater bargain.

In 1974 the Surgeon General of the United States announced national goals for the continuing immunization of America—goals that were projected to be achieved by 1990. That program is not only working, it is ahead of schedule. It is just possible that by 1990, or shortly thereafter, measles and polio will become extinct in the United States—historical curiosities.

It is sad that this splendid program is now threatened by budget cutters and by huge damages awarded against pharmaceutical firms whose vaccines have accidentally caused illness or death. Courts should realize that it is no more possible to manufacture an absolutely safe vaccine than it is to make a faultless car, and the public must accept a minimum of risk in return for the immense benefits; otherwise pharmaceutical firms will give up the manufacture of vaccines.

Further great advances in vaccination are now in the offing. By 1990 most children in the United States should be vaccinated against mumps, measles, and rubella, and measles might soon be eradicated worldwide, just like smallpox. There already exists a vaccine against hepatitis B, a widespread killer, but it costs one hundred dollars a shot. Leprosy still affects about 12 million people. Large-scale clinical trials of vaccines against it are being carried out in Venezuela, India, and Malawi. Research on a vaccine against malaria has become possible since William Trager at Rockefeller University in New York found out how to culture the malaria parasite in human red blood cells, but there are many technical difficulties still to be overcome. The brightest hope lies in genetically engineering the vaccinia (cowpox) virus, now used for vaccination against smallpox, so that it displays on its surface also the markers of other disease-causing organisms, such as the hepatitis B virus and the malaria parasite. A single shot of this genetic composite should be cheap and immunize against a variety of diseases. Cancer of the cervix appears to be generally caused by a papilloma virus, and primary cancer of the liver by hepatitis B virus. Both infections could be prevented by vaccination. For the distant future, the author forecasts vaccination even against some of the other, more common cancers.

Of the 90 million children now born in the world each year, 15 million die during their first year of life, and 6 million of these die from preventable infectious diseases, mostly in the Third World. In addition to the 15 million dead, 180,000 children become lamed by polio. Much more illness could be prevented cheaply by world-wide vaccination, and Desowitz pleads passionately for its being put into effect, because mothers in an African village mourn their children as deeply as mothers do in the Western world. I was much taken by Desowitz's imagined "day in the life of a tropical vacci-nator," a warmhearted story of a Candide of the Third World who sets out from his district health center to vaccinate the children in a distant village against diphtheria, measles, whooping cough, and tetanus. So far only a quarter of the Third World's children have received the combined vaccine, even though it costs only seventy cents per child. Candide's collection excludes smallpox, because that scourge of mankind had already been eradicated by the World Health Organization in what must have been one of the greatest achievements of international cooperation ever. It is quite remark-able that the organization got everyone vaccinated in even the remotest corners of the globe despite the neglect, irresponsibility, incompetence, and corruption that are typical of many countries and frustrate the author's Candide at every step.

Some vaccines may be rendered ineffective by malnutrition, which is widespread, especially among children in the Third World. For undernourished children only the vaccines against smallpox, measles, and polio work, while those against typhoid, mumps, diphtheria, and yellow fever offer no protection. They could be made effective by giving the children a diet high in protein just before and for one or two weeks after vaccination. According to the author, "This modest expedient appears simple enough on paper, but is probably almost impossible in the field, as any despairing administrator of a famine-relief program is well aware." The World Health Organization has laid down minimal nutritional require-ments for people of all ages, but these may be inadequate for many among the undernourished quarter of the world's population, because their intestines are infested by parasites that snatch the food from them and so irritate their bowels that even their minimal rations are poorly absorbed. In the United States, Desowitz writes, "The Third World may be only a few blocks away." The Physicians' Task Force on Hunger in America found children suffering from protein deficiency diseases such as kwashiorkor, which make their tummies swell and their bottoms shrink, diseases that were believed to be confined to the tropics. Other children displayed stunted growth, lethargy, and vitamin deficiencies. These children of the American poor, immunologically impoverished and susceptible to infection, are no longer protected by the government-sponsored immunization programs. According to the task force, the United States requires a famine relief program of its own, which could be provided at a small cost compared with the sums wasted on the useless "Star Wars" program.[1]

When will there be an effective vaccine against AIDS? Desowitz has no answer to this vexing question, but he supplies much interesting and useful information about AIDS. A new disease first came to be suspected in 1979, when a rare form of pneumonia, normally found only in small infants, was diagnosed among five male homosexuals in Los Angeles. Seven years later, 28,000 people in the United States were diagnosed as having AIDS and 270,000 are forecast for 1991, with a total of about 180,000 deaths during the twelve-year period, nearly four times the number of Americans killed in the Vietnam War. The AIDS virus was first identified in 1983 by Luc Montagnier at the Pasteur Institute in Paris and named lymphadenopathy-associated virus (LAV). The author prefers to call it human T-cell leukemia virus (HTLV) III, a name coined by Robert Gallo at the National Institutes of Health in Bethesda, Maryland, in the belief that it is related to the T-cell leukemia viruses. The name now commonly used is human immunodeficiency virus, HIV for short.

Desowitz deals at length with the evidence that this virus really is the cause of AIDS, and that the presence of antibodies against it proves people to be carriers of AIDS. He provides detailed descriptions, not fit for the drawing room, of the habits of male homosexuals. Apparently ten different sexual partners in a day and one hundred in a month used to be common, but a recent drastic fall in the incidence of rectal gonorrhea in San Francisco indicates a drop in promiscuity that may slow the spread of the disease. AIDS is not catching like a cold and is generally transmitted only by blood transfusions, shared syringes, abrasive, mostly homosexual sex, or across the placenta from an infected mother to her embryo, but more rarely by normal monogamous heterosexual intercourse. Against that last generalization Desowitz cites the worrying finding that a small fraction of AIDS patients in Haiti are just such normal monogamous people. Some experts believe AIDS to have been long endemic in Africa and to have remained unnoticed, but the author dismisses this view on the ground that the symptoms are so striking and unique that not even the most obtuse physician could have missed them. However, examinations of frozen samples of human serum taken in central Africa in 1959 have shown one sample to contain antibodies against the AIDS virus, whereas AIDS-positive samples have not been found among American or European samples of serum frozen at that time.

Viruses related to the two known AIDS viruses have been found in African monkeys, which made people wonder if transmission from monkeys to man might have initiated the AIDS epidemic. Desowitz dismisses this possibility on the ground that he has never heard of an African ravishing a monkey, but in a recent issue of *The Lancet* F. Noireau quotes a book by the anthropologist A. Kashamura on the sex habits and cultures of the people near the Great African Lakes: "To rouse a man or a woman to intense sexual activity, one injects into their thighs, their pubic region or their back the blood of a male monkey, for a man, or a female one, for a woman." The book was written before the emergence of AIDS.[2] The description is likely to be authentic because Kashamura is a native of the Great Lakes region and wrote the book about the customs of his own people. Noireau concludes that "these magic practices would constitute an efficient experimental transmission model and could be responsible for the emergence of AIDS in man."[3]

What hope is there of halting the spread of AIDS or saving its victims? According to Desowitz, "we can't look forward to the chemotherapeutic magic bullet pill for succour. Chemotherapy research has yet to gift mankind with any practical effective drug to combat any viral disease—let alone AIDS." In fact a highly effective drug against the herpes family of viruses has been on the market for several years. It is called acyclovir and was discovered in 1977 by Gertrude Elion, Howard Schaeffer, and D. J. Bauer at the Wellcome Research Laboratories. It relieves the pains and prevents the aftermath of shingles, that tormenting disease of the elderly. Drug firms are now synthesizing hundreds of chemical analogues of acy-clovir in the hope that one of them might prove active against AIDS.

Research on all aspects of AIDS is accelerating. The Public Health Service has asked for $351 million in 1987 and $471 million in 1988 for its support. Many good scientists have put their current work aside to study the AIDS virus and its attack on the immune system. In the short space of four years since its discovery, they have deciphered its gene, characterized its mode of replication, provided an inventory of its component molecules, and identified the type of white blood cell that harbors it. These are the essential preliminaries to a rational therapy. Development of a vaccine against AIDS faces the difficulty that the virus mutates rapidly, possibly even to the extent of differing slightly in each individual. Moreover, once a person has been infected, the presence of antibodies against the virus does not necessarily prevent the development of the dis-ease, perhaps because the chromosome of the AIDS virus becomes integrated in the chromosomes of the white blood cells of its host, so that the virus becomes part of the host's own genetic system.

In the United States AIDS research is coordinated by the Na-tional Institutes of Health, who have enlisted some of the best brains in virology, molecular biology, and biochemistry to work on the problem. There is good hope that these outstanding scientists will find ways of preventing the spread of the terrible disease and of saving its victims. In Britain, the Medical Research Council, a gov-ernment agency similar to the NIH, is mobilizing research for vac-cines and drugs against AIDS. Similar efforts are under way in other European countries and will soon become worldwide.

The strength of Desowitz's book springs from his compassion and wide medical experience in many parts of the world. For ex-ample, his description of smokers' emphysema does not end with the analysis of its microscopic manifestations but with the recol-lection of an old friend, the head technician of the London School of Tropical Medicine, who was a chain smoker and succumbed to the disease. His discussion of epidemiology in the tropics goes beyond statistics to the daily problems of poor people's lives. To the individual layman in the United States the book offers much sensible advice about vaccination, diet, and life-style, and to anyone concerned about public health much challenging information about preventable disease. Desowitz tries to make laymen understand how the immune system works at the cellular level, but I was disap-pointed that he stops short of explaining its intriguing molecular mechanism, whose unraveling has been one of the greatest recent triumphs of molecular biology.

The book's weaknesses lie in occasional errors, especially in biochemistry, in a disdain for basic biomedical science that "has not cured anybody" but has in fact been responsible for many of the medical advances described here, such as the Sabin vaccine against polio, and in occasional lapses from the popular to the fatuous, such as Desowitz's final imaginary conversation with a visitor and two colleagues who claim that the conquest of malaria is purely a question of money, or the suggestion that Pasteur might have delayed a crucial experiment because an exasperated Mrs. Pasteur might have said, "What, Louis, you are going to the laboratory today? This is the weekend you promised to clear out the closets." How can any American who has lived in Europe imagine that a woman of the French bourgeoisie of the last century would have dared to ask her husband to demean himself with a task that she would have left to her maid? Even today few Frenchmen will handle a duster. A little farther on, we read that Pasteur was "as frugal as any budget-conscious researcher today." This conjures up a Pasteur counting his research grants, rather than a scientist working in a spartan time when organized state support for science did not yet exist and researchers had to improvise with primitive tools often paid for out of their own pockets. Desowitz writes about "a bacteria"; he would not tell us that he is married to "a women," so why deny the poor bacter*ium* its singular? However, these are minor faults in what may be the first book that explains our natural defenses against infection lucidly to the layman.

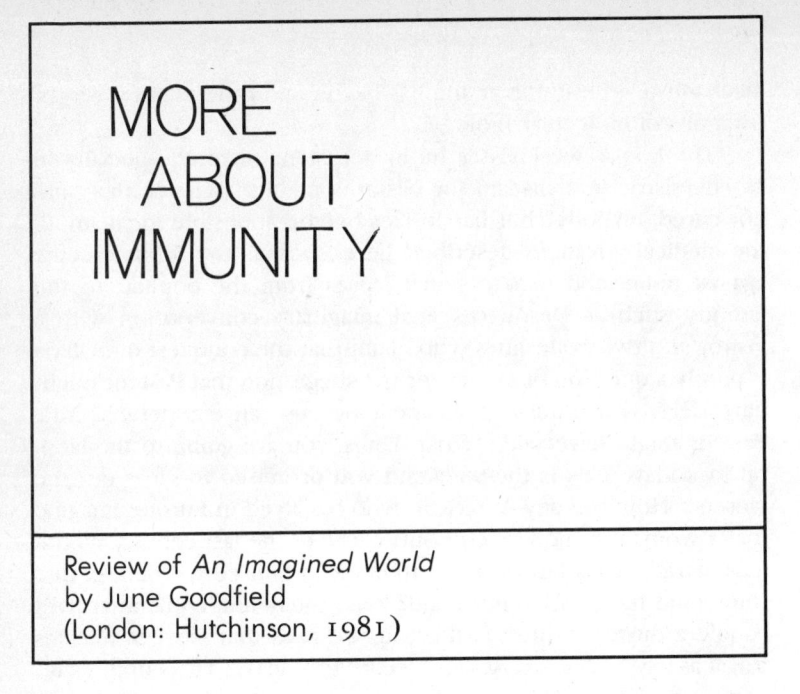

MORE
ABOUT
IMMUNITY

Review of *An Imagined World*
by June Goodfield
(London: Hutchinson, 1981)

"My instinct tells me about this love for things...tells me that I may not manage to cope with people....I suspect I am almost too direct with people. I have too much clarity, and they can't cope with it." Did that failing divert Anna Brito's affections to white blood cells? She personifies Peter Medawar's dictum that a scientist is not a person who merely turns the handle of discovery, "for at every level of endeavour scientific research is a passionate undertaking."[1] When June Goodfield meets Anna at a cancer institute in New York, she is captivated by Anna's imaginative and ardent devotion to her research. She decides to watch Anna and her colleagues at work, as Jane Goodall watched the chimpanzees going about their daily business in the Ugandan forest. To fathom the behavior of that mysterious species *Homo sapiens scientificus*, June Goodfield faithfully observes the scientists' natural habitat, the laboratory, and records their every word, movement, and gesture; she also makes Anna record her thoughts on tape while Anna is not actually being watched. It took Jane Goodall many months of silent, patient waiting until the chimpanzees accepted her as part of the forest and behaved as if she were not there. I wonder if June Good-

234

field's species always acted as if they were unobserved; occasionally one suspects them of declaiming to the gallery.

June Goodfield's heroine is the only daughter of a wealthy Portuguese couple who was expected to cultivate the graces of a girl of her class, to get married, and to have children. Instead, she decided to read medicine. During her clinical years she came to realize that she would not be able to cope with the sight of human suffering; after she had qualified she therefore took up research rather than medical practice. She joined a biological laboratory newly built by the Gulbenkian Foundation and was sent to London to learn immunology.

Anxious to please in a southern way, the Portuguese authorities tell Anna's hosts that they are sending them an experienced research worker who is fluent in English. Faced instead with a mute young girl who has never done an experiment in her life, Anna's supervisor, Dr. Vera, finds her a job in which at least she can't spoil anything. She gives her a microscope and tells her to examine thin sections cut through the spleens and lymph nodes of mice. Some of the mice have had the thymus, a small gland in the neck, surgically removed immediately after birth, and others are normal. Dr. Vera tells Anna to look for differences between the tissues of the two kinds of mouse. June Goodfield explains that the problem is concerned with the development of the animals' defense from infection and perhaps also from cancer. Such defense is the task of white blood cells called lymphocytes, which originate in the bone marrow. Yet, for mysterious reasons, the defense does not function in animals that grow up without a thymus. To find out the thymus's role, Dr. Vera tells Anna to look at the lymph nodes and the spleen, where lymphocytes like to collect.

What is she after? While lymphocytes may normally protect us from cancer, they proliferate in cancers such as leukemia and Hodgkin's disease. Research in lymphocytes may therefore provide clues to the mechanism of immunity and to the causes of cancer. No field could be more important and few are more difficult, both technically and conceptually. Lymphocytes are single cells, like bacteria, but much harder to handle because they are so much more sensitive: they can easily be killed by heat or cold, by too much or too little salt, by bacterial and viral infections, or by wrong nutrients. Their bland appearance under the microscope hides a warren of membranes and organelles that direct their lives. Their surfaces are studded with sensors that tell friend from foe. Their chemistry is a microcosm of unbelievable complexity. When their behavior changes, cause and effect may play blindman's buff with the observer.

Anna withdraws into a corner with Dr. Vera's slides. After a month of staring down a microscope, she is sure that she has discovered something but soon finds that facts do not speak for themselves and, without knowing the language, she cannot speak for them. So, when three months have passed, Dr. Vera tells Anna that it is time for her to go home to Lisbon. Anna's anger brings out the missing words. She had noticed that certain spaces in the spleen that were normally filled with lymphocytes were empty in the spleens of mice without thymuses, even though there were plenty of lymphocytes populating other spaces in their spleens. This made Anna realize that there must be at least two different populations of lymphocyte: both originate in the bone marrow, but while one needs no other organ to mature, the other has to mature in the thymus. She saw that these two different populations home in on different spaces in the spleen. How can lymphocytes tell where they have to go? What traps them when they get there? From now on these questions dominate Anna's thoughts and actions.

After two years in London, Anna returns to Lisbon expecting an eager welcome, but her discovery falls flat among her colleagues there, and her vital questions leave them cold. She sees herself as a dynamic genius wasting among smug mediocrity: she therefore wrenches herself away from home and family and takes up a university lectureship in Glasgow. By now other workers have discovered that immunity needs the cooperation of the two different kinds of lymphocyte. At Glasgow Anna wonders what attracts them to each other. She discovers instead that thymus-derived lymphocytes secrete a factor that prevents them joining up with lymphocytes from the bone marrow. "My discovery is true," she writes to June Goodfield. "I am so excited about it that I am bursting." She composes a suitably sober letter to *Nature* about her find, but when it causes no stir in the world of immunology, she feels like a rejected lover.

Who cares? Why should I believe that it matters? Why do I believe that I carry the responsibility of demonstrating and searching for the evidence of an idea, when I am sure hundreds of other people could do it, and if I died tomorrow it wouldn't matter tuppence? I don't know ... if Jim Watson or Francis Crick ever get depressed. But for the more ordinary of us, the strength that is needed to believe that what you believe in is worth pursuing is very great. [This reviewer has found that even greater strength is sometimes needed to abandon one's beliefs.] ... I thought for the first time of valuable, practical applications if it works. Patients with metastases [spreading cancer cells] have been found to have low numbers of circulating lymphocytes. It is just possible that ... the little things are now trapped in the metastases. If we manage to trace them we would have an early warning detective system of metastases distribution.

These flights of fancy lead Anna to study Hodgkin's disease, the cancer that causes the lymph nodes to swell while there may be a lack of lymphocytes in the blood. Anna wonders if they have got trapped, and if so where, and what is trapping them. To test her ideas, she needs blood, lymph, and tissue samples from patients with Hodgkin's disease. To get them she abandons the security of her lectureship for a precarious research appointment at a cancer institute and hospital in New York. There she tells her biographer,

> Today I have no doubts about ... being right in my thinking. ... I cannot believe the evidence before my eyes ... that Hodgkin's disease may be a form of ecotaxopathy [an imposing term Anna has coined for cells homing in on a target tissue]. ... I am not going to have another thought for four years. Just experiments until the results speak for themselves. ... If people looked at the terribly worried faces of the visitors in the hospital halls ... they would get the feeling of urgency.

Alas, those feelings are not shared by the committees that distribute research grants: her applications are all refused, perhaps because they elicit the same response from the grant committees as the accused elicits from the judge in Heinrich Kleist's comedy *The Broken Jug*: "In your head fact and fancy are kneaded together as in a dough; with every slice you give me some of each."

Hoping for clues to the origins of Hodgkin's disease, Anna starts working with a Chinese woman physician who has devoted many years to the recording of all aspects of that disease in over 250 patients. Anna spends weeks browsing through the records. After a marathon session lasting from early one morning until late the following night, we hear her cry "Eureka," because she believes that the records point to a simple answer. The count of lymphocytes in the patients' blood goes up just before its content of iron goes down. What is more, the patients' lymphocyte surface carries a protein that stores iron. Iron must be the key!

Anna's hunch makes her search for signs of faulty iron metabolism in patients' spleens. She argues that this malfunction should manifest itself by a fluorescence when the cells are viewed under ultraviolet light. Anna finds it. No one else had noticed it, she says, because no one had looked, and no one had looked because no one had conceived the right theory. Connections between iron, Hodgkin's disease, and other forms of cancer now pop up wherever she turns. Before puberty the incidence of Hodgkin's disease is the same for both sexes, but afterward it becomes more frequent for boys, clearly because iron is lost by menstruation in girls. Anna reads of a multiple incidence of Hodgkin's disease in Sheffield: obviously because Sheffield is full of steel mills. She finds the spleens of patients with cancer of the lymph choked with iron. Anna's colleagues have discovered a factor that stimulates white blood cells to form colonies, especially in leukemic patients. They have also found an antagonist that prevents formation of colonies. Anna has a hunch that this might be a protein that binds iron tightly. Her idea turns out to be right. Here at last she has found the answer to the question raised by her first piece of research. What directs the traffic of lymphocytes? Iron.

"I have been spending 24 hours a day just thinking," she tells her biographer. (What about her resolve to stop thinking for four years and stick to experiments?) "If all this is true, then ... in leukaemia ... you have increased exposure to one type of iron ... so you can expect ... a lot of iron in the blood. That was my prediction and today it came true ... we have made a significant step forward in our understanding of leukaemia. I don't suppose anyone else will believe me but I do." This reminded me of a letter to my father-in-law, written in 1949, which I recently found. In it I announced with loud fanfares that I had solved the structure of hemoglobin, the protein of red blood cells, a problem on which I had worked since 1937. A few months later Francis Crick proved to me that my solution was nonsense.

Are Anna's ideas right? Those who have spent a lifetime studying Hodgkin's disease are skeptical. If iron were important for Hodgkin's disease, the director of Anna's institute tells her gently, then changes in patients' diet should affect its course, but they do not. There should be an association between the disease and certain genetic disorders affecting iron metabolism, but none has been found. One day, Dr. Henry Kaplan, a world authority on Hodgkin's disease, visits Anna's laboratory. He says to her, "Maybe the avidity for iron that you are finding is just one more thing that cells do when they are activated, a secondary reaction and not the primary cause. I think that your study of iron and iron-binding proteins is wonderful, but don't walk too rapidly towards a particular set of molecules. You will find this a can of worms." But Anna is deaf to their well-founded objections and adheres to her faith in Hodgkin's disease being primarily a defect in the white blood cells' handling of iron—rather as another iron lady sticks to monetarism despite all proofs that it is making the patient worse. Others must have pointed out to Anna that if the disease were primarily caused by an error in iron metabolism, rather than by proliferation of the descendants of a single mutated cell, or cells made cancerous by an as yet unknown virus, as is widely believed, then susceptibility to the disease should be inherited. This is not the case.

Anna says: "I long to be old, and I resent . . . that your ideas may not be worth . . . a penny before you are fifty." In fact, cancer theories constitute an occupational disease among elderly Nobel laureates, and few scientists take them seriously. Otto Warburg, perhaps the greatest German biochemist ever, believed to his death that cancer cells can get their energy without oxygen and are therefore different from normal cells. One day I visited him in his Institute for Cell Research, an elegant Rococo castle on the outskirts of Berlin. On entering, I was faced by a life-size marble bust of Otto Warburg. The real Warburg then received me affably in his library. Soon his talk turned to his cancer theory: "The other day I lectured about it, and do you know what happened? A student got up afterwards and contradicted me. A student contradicted me, a Nobel laureate! In the old days, this could never have happened. Willstätter [a famous chemist] would have crushed him with an angry glance." Is this the position Anna craves for? One great American chemist now believes that massive doses of vitamin C prolong the lives of cancer patients. Even more dangerous are physicians who believe in cancer cures. I remember one who persuaded himself and others that tumors regress in response to a treatment he devised. He became head of a cancer ward where he was all-powerful enough to inflict his treatment on his hapless patients for over two decades— even though it merely increased their pains without having any therapeutic effect on their tumors. The anxious public and the media invariably acclaim such people and denounce skeptics for closing their minds to unorthodox ideas. I am worried that Anna is on the slippery path to that doubtful company and that June Goodfield might become her publicity agent.

So much for the plot. June Goodfield's account of it is lucid and lively, and it makes the reader fond of her heroine, even though at times she sounds a little naïve. "I've never been like this in my whole life," she records on her tape. "For the first time I'm really frightened . . . frightened of making a mistake, of being wrong, and that is terrible, it is just terrible." I have learned from people who have sailed round the world single-handed that they were dead scared of making a mistake all the way. It was that constant fear that kept them on course. Another time, Anna explains, "I'm paralysed . . . just paralysed. For one thing the virus is going to be the biggest red herring in the history of cancer. And everyone in the house of science is just standing in the corner facing the wall. Here we've been really working for six weeks and we've seen something. And do you know why . . . ? Because we've *thought*." But thought is not enough. You also need judgment. Thirty years ago a brash young American crystallographer had thought and then went around saying: "Give me a million dollars and I'll solve the structure of proteins in five years." He found a gullible philanthropist who gave him the million, but he never solved any protein structure because his thought was wrong.

Anna declares, "The talks I gave were good, but the most important thing was the perspective I gained into myself as the woman I now am and into the exact significance of our finds in an evolutionary perspective. . . . The ecstasy comes from the realisation that the process of growth — whether a cancer or benign growth — is an ancestral one, fallen from meteorites." Poof! What pretentious trash! "It's going to be so exciting, the next 20 years" sounds like a teenager after her first audition onstage.

Despite such pompous platitudes, the book is an exciting account of a romantic, gifted, and determined young woman's attempt to solve one of the basic problems of medicine. Even though her theories are probably incorrect, they have stimulated her to discover new facets of lymphocyte behavior and chemistry. I was a little put off by her frantic and somewhat obsessive style of doing research, but anyone wanting a vivid day-to-day account of the hopes, the tedium, the triumphs, and disappointments of research in a cancer laboratory will find it good reading. June Goodfield substitutes simple words for esoteric terms and explains the science accurately and clearly to the layman.

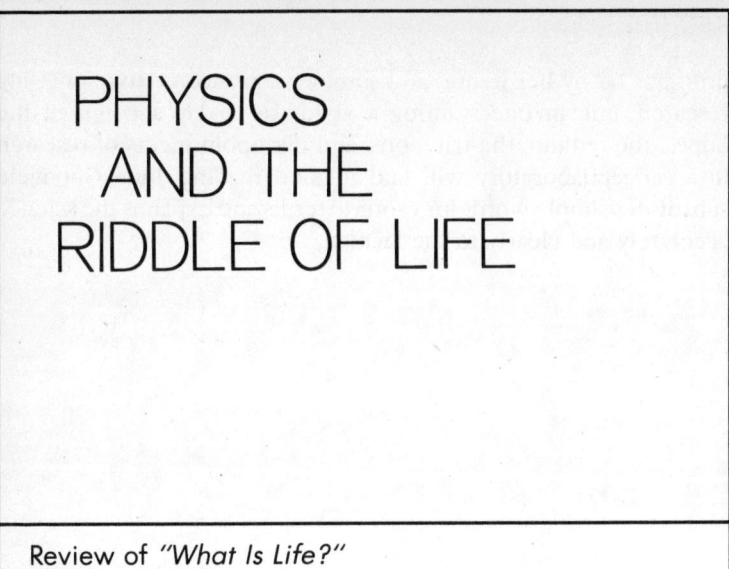

PHYSICS
AND THE
RIDDLE OF LIFE

Review of *"What Is Life?"*
and Molecular Biology by Erwin Schrödinger
(Cambridge: Cambridge University Press, 1987)

In the early 1940s Erwin Schrödinger, the discoverer of wave mechanics, worked at the Institute for Advanced Studies in Dublin. One day he met P. P. Ewald, another German theoretician who was then professor at the University of Belfast. Ewald, who had been a student in Göttingen before the First World War, gave Schrödinger a paper that had been published in the *Nachrichten aus der Biologie der Gesellschaft der Wissenschaften* in Göttingen in 1935. It was by N. W. Timoféeff-Ressovsky, K. G. Zimmer, and Max Delbrück and was titled "The Nature of Genetic Mutations and the Structure of the Gene."[1] Apparently Schrödinger had been interested in that subject for some time, but the paper fascinated him so much that he made it the basis of a series of lectures at Trinity College, Dublin, in February 1943; they were published by Cambridge University Press in the following year, under the title *What Is Life? The Physical Aspect of the Living Cell.*

The book is written in an engaging, lively, almost poetic style ("The probable life time of a radioactive atom is less predictable than that of a healthy sparrow"). It aroused wide interest, especially among young physicists. Up to 1948 it drew sixty-five reviews, and it has probably by now sold 100,000 copies. It has since become a classic that has provided a nourishing habitat for historians, sociologists, and philosophers of science who have commented on it, on the comments on it, or on the comments on the comments on it. A Ph.D. thesis published on the subject in 1979 contains over 120 references, excluding the sixty-five reviews.[2] François Jacob has explained the reasons for the book's impact best:

> After the war, many young physicists were disgusted by the military use that had been made of atomic energy. Moreover, some of them had wearied of the turn experimental physics had taken . . . of the complexity imposed by the use of big machines. They saw in it the end of a science and looked around for other activities. Some looked to biology with a mixture of diffidence and hope. Diffidence because they had about living beings only the vague notions of the zoology and botany they remembered from school. Hope, because the most famous of their elders had painted biology as full of promise. Niels Bohr saw it as the source of new laws of physics. So did Schrödinger, who foretold revival and exaltation to those entering biology, especially the domain of genetics. To hear one of the fathers of quantum mechanics ask himself: "What is Life?" and to describe heredity in terms of molecular structure, of interatomic bonds, of thermodynamic stability, sufficed to draw towards biology the enthusiasm of young physicists and *to confer on them a certain legitimacy* [The emphasis is mine]. Their ambitions and their interests were confined to a single problem: the physical nature of the genetic information.[3]

Ilya Prigogine found that Schrödinger's book was an inspiration for his work on nonequilibrium thermodynamics; Seymour Benzer, Maurice Wilkins, and Gunther Stent have said that the book was decisive in drawing them from physics into biology; Francis Crick told me that he found it interesting but would have switched to biology anyway. I was already in the thick of trying to solve the structure of proteins when it was published, and I may have been encouraged by its quotation of C. D. Darlington's view that genes are made of protein. Crick wrote in 1965,

> On those who came into the subject just after the 1939–1945 war Schrödinger's little book . . . seems to have been peculiarly influential. Its main point—that biology needs the stability of

chemical bonds and that only quantum mechanics can explain this—was one that only a physicist would feel it necessary to make, but the book was extremely well written and conveyed in an exciting way the idea that, in biology, molecular explanations would not only be extremely important but also that they were just around the corner. This had been said before, but Schrödinger's book was very timely and attracted people who might otherwise not have entered biology at all.[4]

However, in 1971 Crick added, "I cannot recall any occasion when Jim Watson and I discussed the limitations of Schrödinger's book. I think the main reason for this is that we were strongly influenced by Pauling, who had essentially the correct set of ideas. We therefore never wasted any time discussing whether we should think in the way Schrödinger did or the way Pauling did. It seemed quite obvious to us we should follow Pauling."[5]

Neither can I recall Crick, Watson, John Kendrew, and me ever discussing the bearing of Schrödinger's book on structural molecular biology during our years together at the Cavendish Laboratory; Stanley Cohen wrote that few of the many scientists participating in Delbrück's phage course at Cold Spring Harbor in 1944 had read Schrödinger, "and in all the social and intellectual activities of these [postwar] summers I do not recall any mention of Schrödinger."[6] The participants included such pioneers of molecular genetics and biochemistry as Salvador Luria, Alfred Hershey, André Lwoff, Jacques Monod, and Jean Brachet. Hence the book does not appear to have had much impact on the people already working in the field.

Schrödinger's book is written for the layman and begins with the chapter "The Classical Physicist's Approach to the Subject." He asks how events in space and time taking place in a living organism can be accounted for by physics and chemistry. "Enough is known about the material structure of life to tell exactly why present-day physics cannot account for life. That difference lies in the statistical point of view. It is well-nigh unthinkable that the laws and regulations thus discovered [i.e., by physics] should apply immediately to the behaviour of systems which do not exhibit the structure on which these laws and regularities are based." Schrödinger jumps to this conclusion after reading that genes are specific molecules of which each cell generally contains no more than two copies. He had entered Vienna University in 1906, the year that Ludwig Boltzmann died, and had been taught physics by Boltzmann's pupils. He remained deeply influenced by Boltzmann's thoughts throughout his life. According to Boltzmann's statistical thermodynamics, the behavior of single molecules is unpredictable; only the behavior of large numbers is predictable. In genetics, therefore, Schrödinger concludes, "We are faced with a mechanism entirely different from the probabilistic ones of physics." This difference forms the guiding theme of his book.

In the first chapter, Schrödinger illustrates the meaning of statistical thermodynamics by the examples of Curie's law, of Brownian motion and diffusion, and of the $\sqrt{n}$ rule. His next two chapters, on the hereditary mechanisms and on mutations, give brief popular introductions to textbook knowledge on these subjects available at the time. They reveal one vital misconception in Schrödinger's mind: "Chromosomes," he writes, "are both the law code and the executive power of the living cell." In fact, biochemists had shown that the executive power resides in enzyme catalysts, and in 1941 G. W. Beadle and E. L. Tatum discovered that single genes determine single enzymatic activities;[7] that discovery led to the one-gene, one-enzyme hypothesis, an idea that had already been foreshadowed by the Cambridge biochemist and geneticist J. B. S. Haldane,[8] and that has become central to an understanding of biology. Schrödinger does not appear to have heard of this.

The next two chapters form the backbone of his book; they are called "The Quantum-Mechanical Evidence" and "Delbrück's Model Discussed and Tested" and, as C. H. Waddington first spotted, they are largely paraphrased versions of the paper by Timoféeff, Zimmer, and Delbrück.[9] That paper covers fifty-five pages and is divided into four sections. The first section, by Timoféeff, describes the mutagenic effects of X rays and gamma rays on the fruit fly *Drosophila melanogaster*. He shows that the spontaneous mutation rate of the fly is low and that it is raised about fivefold by a rise in temperature of $10°$ C. Ionizing radiations increase that rate as a linear function of the dose, independent of its time distribution, of the wavelength, and of the temperature during irradiation.

The second section of the paper is by Zimmer and applies the target theory to Timoféeff's results. The number of mutations $x = a(1 - e^{-kD})$, where a and k are constants and D is the dose. Zimmer next asks whether the mutations had arisen by the direct absorption of quanta, by the passage of secondary electrons through a sensitive volume, or by the generation of ion pairs. If the dose is measured in roentgens, the number of quanta required to produce a given dose diminishes with diminishing wavelength. Thus, direct absorption of quanta is inconsistent with the linear dependence of the mutation rate on the dose. The same applies to secondary electrons. Only the number of ion pairs is proportional to the dose, obviously, since that is how the dose is measured. Zimmer therefore concludes that a single hit suffices for the production of one mutation and that this hit consists of either the formation of an ion pair or a transition to higher energy.

The third section of the paper is by Delbrück and bears the title *"Atomphysikalisches Modell der Mutation"* ("A Model of Genetic Mutation Based on Atomic Physics"). Delbrück reminds us that the concept of the gene began as an abstract one, independent of physics and chemistry, until it was linked to chromosomes and later to parts of chromosomes, which were estimated to be of molecular size. Since he and his colleagues had no means of discovering the chemical nature of genes directly, they attacked the problem indirectly by studying the nature and the limits of their stability and by asking if these were consistent with the knowledge that atomic theory has provided about the behavior of well-defined assemblies of atoms.

Such assemblies can undergo discrete and spontaneous transitions of vibrational and electronic states. Vibrational transitions are very frequent and involve no chemical changes. From electronic transitions the assemblies may either revert to the ground state or reach a new equilibrium state after undergoing an atomic rearrangement, for example to a tautomeric form. The fivefold rise in spontaneous mutation frequency for a $10°C$ rise in temperature leads Delbrück to derive an activation energy of about $1.5eV$ and an average lifetime of a few years, when half the molecules composing the gene will have undergone an electronic transition.

Delbrück then describes how on the average X rays lose energy to secondary electrons in portions of $30eV$ per ionization, which is $1000 \times kT$ and 20 times the energy of activation of $1.5eV$ needed for a spontaneous mutation. However, to produce as much as $1.5eV$, the ionization must not occur too far away from its target. We knew too little about the ways in which the energy of photoelectrons is dissipated to determine the absolute value of the dose needed to induce a mutation with a probability of unity, but that dose, expressed as the number of ionizations per unit volume, was likely to be about ten to one hundred times smaller than the number of atoms of the gene per unit volume. Delbrück now calculates that dose as follows.

A frequently observed X-ray mutation (eosin) occurs with a dose of 6,000 roentgens once in 7,000 gametes. Hence a probability of unity of its occurrence needs a dose of 42×10^6 roentgens. One roentgen produces about 2×10^{12} ion pairs in 1 milliliter of water, whence 42×10^6 roentgens produce about 10^{20} ion pairs. Since 1 milliliter of water contains about 10^{23} atoms, this means that at least 1 in 1,000 atoms becomes ionized. However, Delbrück cautiously refrains from concluding that a gene is likely to consist of a thousand atoms.

Schrödinger used Delbrück's result to point out "that there is a fair chance of producing a mutation when an ionization occurs not more than about 10 atoms away from a particular spot on the chromosome," but research published while Schrödinger was writing his book showed such calculations to be meaningless. In a paper that appeared in *Nature* in June 1944, Joseph Weiss pointed out that the biological effects of ionizing radiation are caused principally by the generation of hydroxyl radicals and hydrogen atoms in the surrounding water.[10] E. Collinson, F. S. Dainton, D. R. Smith, and S. Tazuke,[11] and independently G. Czapski and H. A. Schwartz,[12] later discovered that the supposed hydrogen atoms were in fact hydrated electrons.[13] Hydroxyl ions and hydrated electrons have half-lives of about 1.0 millisecond (assuming a concentration of 1 micromole H_2O_2) and of 0.5 millisecond respectively, in which times they can diffuse to their targets even if they are generated more than a thousand atomic diameters away from them.

Delbrück concludes that it is premature to make the description of the gene any more concrete than the following:

> We leave open the question whether the single gene is a polymeric entity that arises by the repetition of identical atomic structures or whether such periodicity is absent; and whether individual genes are separate atomic assemblies or largely autonomous parts of a large structure, i.e. whether a chromosome contains a row of separate genes like a string of pearls, or a physico-chemical continuum.

I found the Timoféeff, Zimmer, and Delbrück paper, and especially Delbrück's part, most impressive. Delbrück was a theoretical physicist whose interest in biology had been aroused by Niels Bohr's lecture "Light and Life," delivered in Copenhagen in 1932. In that lecture Bohr had said,

> The existence of life must be considered as an elementary fact that cannot be explained, but must be taken as a starting point in biology, in a similar way as the quantum of action, which appears as an irrational element from the point of view of classical mechanical physics, taken together with the existence of elementary particles, forms the foundation of atomic physics. The asserted impossibility of a physical or chemical explanation of the function peculiar to life would be . . . analogous to the insufficiency of the mechanical analysis for the understanding of the stability of atoms.[14]

The search for Bohr's elementary fact of life had fired Delbrück's imagination. He was only twenty-nine years old, working as assistant to Otto Hahn and Lise Meitner in the Kaiser Wilhelm Institut für Chemie in Berlin and doing his biological work as a sideline, but his paper shows the maturity, judgment, and breadth of knowledge of someone who had been in the field for years. It is imaginative and sober, and its carefully worded predictions have stood the test of time. The paper won him a Rockefeller fellowship to Pasadena to work with the *Drosophila* geneticist T. H. Morgan. There he met Linus Pauling, with whom he published an important paper in 1940. That paper was an attack on the German theoretician Pascual Jordan, who had advanced the idea that there exists a quantum-mechanical stabilizing interaction, operating preferentially between identical or near-identical molecules, which is important in biological processes such as the reproduction of genes. Pauling and Delbrück pointed out that interactions between molecules were now rather well understood and give stability to two molecules of *complementary* structure in juxtaposition, rather than to two molecules with necessarily *identical* structures. Complementariness should be given primary consideration in the discussion of the specific attraction between molecules and their enzymatic synthesis.[15] In 1937 the Cambridge geneticist and biochemist J. B. S. Haldane had made a similar suggestion: "We could conceive of a [copying] process [of the gene] analogous to the copying of a gramophone record by the intermediation of a negative, perhaps related to the original as an antibody is to an antigen."[16] Schrödinger mentions neither of these important ideas.

Schrödinger's last two chapters do contain his own thoughts on the nature of life. In "Order, Disorder, and Entropy" he argues that "the living organism seems to be a macroscopic system which in part of its behaviour approaches to that purely mechanical (as contrasted with thermodynamical) conduct to which all systems tend, as the temperature approaches the absolute zero and the molecular disorder is removed." He comes to this strange conclusion on the ground that living systems do not come to thermodynamic equilibrium, defined as the state of maximum entropy. They avoid doing so, according to Schrödinger, by feeding on negative entropy. I suspect that Schrödinger got that idea from a lecture by Ludwig Boltzmann on the second law, delivered before the Imperial Austrian Academy of Sciences in 1886:

> Hence the general battle for existence of living organisms is not one for the basic substances—these substances are abundant in the air, in water and on the ground—also not for energy that every body contains abundantly, though unfortunately in a non-available form, but for entropy which becomes available by the transition of energy from the hot sun to the cold earth.[17]

Franz (later Sir Francis) Simon, then at Oxford, pointed out to Schrödinger that we do not live on $-T\Delta S$ alone but on free energy.[18] Schrödinger deals with that objection in the second edition of his book; he writes that he had realized the importance of free energy but had regarded it as too difficult a term for his lay audience; to me this seems a strange argument, since the meaning of entropy is surely harder to grasp. Schrödinger's postscript did not satisfy Simon, who pointed out to him in a letter that "the reactions in the living body are only partly reversible and consequently heat is developed of which we have to get rid to the surroundings. With this irreversibly produced heat also flow small amounts (either $+$ or $-$) of reversibly produced heat ($T\Delta S$), but they are quite insignificant and therefore cannot have the important effects on life processes which you assign to them."[19]

In fact, it was known when Schrödinger wrote his book that the primary currency of chemical energy in the cells is ATP, and that the free energy stored in ATP is predominantly enthalpic. Prigogine disagrees with Simon's and my objections and explains that organisms in a steady state liberate as much positive entropy as the negative entropy they absorb. I find this argument hard to follow, because plants absorb free energy only in the form of radiation, which they use to create order from disorder. In other words, they convert enthalpy into negative entropy.

The final chapter, "Is Life Based on the Laws of Physics?", re-iterates and amplifies the central argument already stated at the beginning of the book. According to Delbrück, Schrödinger writes, the gene is a molecule, but the bond energies in molecules are of the same order as the energy between atoms in solids, for example, in crystals, where the same pattern is repeated periodically in three dimensions and where there exists a continuity of chemical bonds extending over large distances. This leads him to the famous hypothesis that the gene is a linear one-dimensional crystal, simply lacking a periodic repeat: an aperiodic crystal. A single such crystal, or a pair of them, directs the orderly process of life. Yet, according to Boltzmann's laws, their behavior must be unpredictably erratic. Schrödinger concludes that "we are faced with a mechanism entirely different from the probabilistic one of physics, one that cannot be reduced to the ordinary laws of physics, not on the ground that there is any 'new force' directing the behaviour of single atoms within an organism, but because the construction is different from any yet tested in the physical laboratory."

I wonder why Schrödinger did not adhere to Delbrück's much better formulation of "a polymeric entity that arises by the repetition of identical atomic structures." One could argue over the distinction between aperiodic and identical, but Delbrück could not have meant structures that are *completely* identical, since these could contain no information. Schrödinger does suggest that the genetic information might take the form of a linear code, analogous to the Morse code.

He argues that the nature of the gene allows only one general conclusion: "Living matter, while not eluding the laws of physics as established to date, is likely to involve other laws of physics hitherto unknown which, however, once they have been revealed, will form as integral a part of this science as the former." Schrödinger is thus drawn to the same conclusion that Niels Bohr had been, apparently unknown to Schrödinger, twelve years earlier, and one that young physicists found equally inspiring.

Schrödinger next refers to a paper by Max Planck, "Dynamical and Statistical Laws." Dynamical laws control large-scale events such as the motions of the planets or of clocks. Clockworks function dynamically, because they are made of solids kept in shape by London-Heitler forces, strong enough to elude disorderly heat motions at ordinary temperatures. An organism is like a clockwork in that it also hinges on a solid: the aperiodic crystal forming the hereditary substance, largely withdrawn from the disorder of heat motion. The single cog of this clockwork is not of coarse human make but is the finest piece ever achieved along the lines of the Lord's quantum mechanics. C. D. Darlington at Oxford had advised Schrödinger that genes are likely to be protein molecules, as was then generally believed; Schrödinger quotes that information but does not mention that proteins are long chain polymers made up of some twenty different links that might form the kind of aperiodic patterns or linear code he had in mind. He must also have been unaware that the true chemical nature of that "finest piece" was actually published while he was writing his book. In January 1944 there appeared in the *Journal of Experimental Medicine* a paper by O. T. Avery, C. M. McLeod, and Maclyn McCarty that reported conclusive evidence that genes are made not of protein but of DNA.[20] In the fullness of time, that discovery has led the majority of scientists to the recognition that life can be explained on the basis of the existing laws of physics.

The apparent contradictions between life and the statistical laws of physics can be resolved by invoking a science largely ignored by Schrödinger. That science is chemistry. When Schrödinger wrote "The regular course of events, governed by the laws of physics, is never the consequence of one well-ordered configuration of atoms, not unless that configuration repeats itself many times," he failed to realize that this is exactly how chemical catalysts work. Given a source of free energy, a well-ordered configuration of atoms in a single molecule of an enzyme catalyst can direct the formation of an ordered stereospecific compound at a rate of 10^3 to 10^5 molecules a second, thus creating order from disorder at the ultimate expense of solar energy. Haldane pointed this out in 1945, in his review of Schrödinger's book.[21]

Chemists could also have told him that there is no problem in explaining the stability of polymers that living matter is made of, because their bond energies range from $3eV$ upward, corresponding to a half-life for each bond of at least 10^{30} years at room temperature.

The difficulty resides in explaining how their aperiodic patterns are accurately reproduced in each generation. There is no mention of this central problem in Schrödinger's book.

Research has cleared away the apparent contradiction between the randomness of single molecular events and the orderliness of life that exercised Schrödinger. The orderliness depends on fidelity of reproduction of the genetic message every time a cell divides, and the fidelity of protein synthesis. The genetic message is encoded in a sequence of nucleotide bases along a chain of DNA. That chain is paired to another carrying a complementary sequence of bases. The two chains are coiled around each other in a double helix in which each adenine (A) forms two hydrogen bonds with a thymine (T) and each guanine (G) forms three hydrogen bonds with a cytidine (C). At cell division the two strands of the parent double helix separate, and each forms a template for the formation of a new complementary strand, resulting in two daughter double helices with the same base sequence as the parent double helix. The necessary monomers are supplied in the form of nucleoside triphosphates, which carry the energy for the formation of the growing chain in the form of an energy-rich phosphorus-oxygen bond; the synthesis of new chain links is catalyzed by an enzyme or system of enzymes that attach themselves to the end of the double helix, unwind it, hold each parent strand rigidly in the conformation needed to catalyze the formation of a new chain link, move forward one step, catalyze the formation of the next link, and so on. Arthur Kornberg and his colleagues at Stanford University have worked out how these enzymes function in *Escherichia coli.*[22]

How do they ensure that at each step of elongation only the nucleotide complementary to that on the parent strand is linked to the daughter strand? On the one hand, chemical kinetics tell us that the four alternative trinucleotides must be bombarding the active site of the enzyme at the diffusion rate of about 10^9 molecules a second. On the other hand, their rates of dissociation from the active site vary, depending on their ability to form complementary hydrogen bonds with the base of the parent strand. Only if the incoming nucleoside triphosphate is oriented correctly in the active site of the enzyme and if the hydrogen bonding groups of its base are complementary to those of the parent base will the new nucleotide remain in the active site long enough for a new chain link to be formed.

As Delbrück foresaw, the main source of spontaneous mutations is not the cleavage of covalent bonds in the parent strand. One source was believed to be the existence of tautomeric forms of the bases, which differ in their arrangement of hydrogen bond donor and acceptor groups. Such changes allow a G to pair with a T, or a C with an A, or a G with an A, but in fact such mispairing probably occurs by another mechanism at apparently less cost in free energy. X-ray analysis of synthetic oligonucleotides has shown that mismatched bases can form hydrogen bonds with each other and be incorporated in the double helix with only slight distortions of the bond angles in the phosphate ester chain. Finally, a G-A pair can form, again with only minor distortions of the double helix, if either of the two bases is inverted about its bond to the ribose.[23] Judging by the frequency of these mistakes, the error rate in the reproduction of the genetic message should be 10^{-4} to 10^{-5} per nucleotide; in fact the measured error rate in *E. coli* is 10^{-8} to 10^{-10}, at least three orders of magnitude less than the theoretically expected one.

How does Nature defeat statistical thermodynamics? One of its tricks is a proofreading and editing mechanism unraveled initially by Kornberg and others at Stanford and subsequently by A. R. Fersht and his colleagues, first at Stanford and later in London at Imperial College.[24] In *E. coli* the enzyme that catalyzes the elongation of the DNA chain has a "second look" at the base pair just joined to the daughter double helix; it excises wrongly paired bases, then incorporates the correct ones. Proofreading and editing, however, are also subject to the errors imposed by chemical kinetics, which means that once in a thousand times, say, the correct base is excised and must then be reincorporated in the growing chain. This costs energy. If proofreading is too rigorous, too much energy is wasted in excising and reincorporating correctly paired bases; if it is not rigorous enough, too many copying errors are left uncorrected.

Using bacterial mutants that are either exceptionally error-prone or error-free, Fersht has measured the cost of fidelity and shown how Nature achieves the best compromise by increasing fidelity by two orders of magnitude to about 5×10^{-7}, but still not enough to account for the observed mutation frequency of only 10^{-8} to 10^{-10} in the replication of *E. coli.*[25] B. W. Glickman and Miroslav Radman have discovered a second proofreading mechanism that can distinguish the parent strand from the daughter strand by virtue of the fact that some of the parent strand's bases have become methylated, while those of the daughter strand are still bare. When the mechanism finds a mismatched base pair in the daughter strand, it excises it and replaces it with the correct one, thus reducing the error rate by the missing one or two orders of magnitude.[26] The error rate of a viral RNA by an enzyme that is incapable of repairing mismatches was found to be about 10^{-4} per doubling, a quite unacceptably high rate even for a bacterium.[27]

The genetic message's function is to code for the sequence of amino acids along protein chains, but DNA does not code for proteins directly. Instead, the genetic message is first transcribed into messenger RNA and then translated into a sequence of amino acids in a protein chain. If enzymes are to work effectively, mistakes in the sequence must be infrequent. Transcription of DNA into RNA is not subjected to proofreading and excision repair, perhaps because messenger RNAs are rarely longer than 10^4 base pairs, so that greater error rates are acceptable in them than in DNA replication. However, translation of RNA into protein presents problems that were first pointed out by Linus Pauling, characteristically, *before* the enzymatic machinery for protein synthesis was unraveled.[28] Certain pairs of amino acids differ by only one methyl group.

It is easy to imagine an active site of an enzyme that efficiently rejects an amino acid that is a misfit because it is too large by one methyl group, but it is hard to see how an active site could discriminate well against an amino acid that merely leaves a hole, because it is too short by one methyl group. The ratio of the reaction rates v of two amino acids A and B whose side chains differ in length by a single methyl group, one just fitting the active site and the other being too short, is given by the equation

$$v_A/v_B = \frac{[A]}{[B]} e^{-\Delta G_b /RT},$$

where ΔG_b is the difference in Gibbs binding energy resulting from the contribution of the side chain. ΔG_b is not likely to be more than $3k$cal mol^{-1}. If $[A] = [B]$ this means that $v_A/v_B < 200$, implying an error rate greater than 0.5 percent.[29] Yet when R. B. Loftfield and D. Vanderjagt tried to measure the error rate in such a situation, they found it to be only 3 parts in 10,000 and concluded "that the precision . . . of peptide assembly is very great, much greater than can be deduced from the study of non-biological chemical reactions."[30]

Fersht showed how this low error rate is achieved without making Boltzmann turn in his grave.[31] Nature makes use of the fact that the selection of the correct amino acid into the growing protein chain proceeds in two stages, both catalyzed by the same enzyme. In the first stage the amino acid is coupled to a phosphate to give it an energy-rich bond; in the second stage it is transferred to an adapter, a molecule of RNA that carries the anticodon triplet of nucleotide bases complementary to the coding triplet for that particular amino acid. At the first stage of the reaction, the enzyme rejects amino acids with side chains that are misfits in the active site because they are too long but reacts with amino acids whose side chains are too short with the large error rates predicted by Pauling. The second stage of the reaction takes place at a different active site of the same enzyme. It is constructed so as to fit those amino acids that were too short for the previous active site. It cleaves them from the adapter RNA and sets them free some hundreds of times faster than the correct amino acid. This second stage can thus reduce the error by a further two orders of magnitude, giving a total error rate of only 10^{-4}. Fersht calls this a double sieve mechanism: the first sieve rejects the amino acids that are too large; the second rejects the ones that are too small. A further stage of editing may reduce possible errors in the recognition by the coding triplet on the messenger RNA of the anticodon triplet on the adaptor RNA, thus ensuring incorporation of the correct amino acid into the growing protein chain.[32]

We can see that life has resolved the apparent conflict between the unpredictable behavior of single molecules and the need for order by making enzymes sufficiently large to stabilize them in unique structures, capable of immobilizing other molecules in their active sites and bringing them into juxtaposition so that they can react at high rates. Yet enzymes are long chain polymers. What makes their chains fold up to form unique and largely rigid structures when entropy drives them to form random coils? X-ray analysis has shown the interior of proteins to be closely packed jigsaws of amino acids with hydrocarbon side chains that adhere to each other. They do so partly by dispersion forces, which are enthalpic, and partly thanks to the entropy that is gained by the exclusion of water from the protein interior. When the protein chain achieves maximum entropy by unfolding to a random coil, both polar and nonpolar groups are exposed to water, which adheres to them and becomes immobilized so that its entropy is diminished. When the chain folds up to its unique structure, the polar groups on the main chain form hydrogen bonds with each other, the nonpolar side chains pack together, and the bound water molecules are set free. The resulting gain in translational and rotational entropy of the water more than compensates for the loss of rotational entropy of the protein chain. Thus it is the water molecules' anarchic distaste for the orderly regimentation imposed on them by the unfolded protein chain that provides a major part of the stabilizing free energy of the folded one and keeps it in its unique, enzymatically active structure.

I feel that I should not close this story without telling what became of the scientists whose paper Schrödinger popularized.

Delbrück, whose Rockefeller fellowship had taken him to Pasadena, stayed there, with short interruptions, for the remainder of his life. In the early 1940s he founded bacteriophage genetics and later, with Salvador Luria, bacterial genetics; he became the leader of an enthusiastic band of young people who developed these new fields of research. In 1969 he, Luria, and A. D. Hershey received the Nobel Prize in physiology or medicine "for their discoveries concerning the replication mechanism and the genetic structure of viruses." Delbrück died in Pasadena in 1981.

While Delbrück's life was happy, Timoféeff's life seems tragic to me, even though I am told that he did not regard it as such. He began his research on *Drosophila* in Moscow in the early 1920s. According to Zhores Medvedev, "In 1924, the Soviet Government made a special exchange agreement with Germany. The famous Kaiser Wilhelm Institute for Brain Research in Berlin-Buch was invited to help organize in Moscow a laboratory for brain research, specially designed to study the brain of Lenin, who had died in January 1924. (At the time of his death, Lenin was considered to be the greatest of geniuses, and his brain was expected to be unique.)"[33]

In a lecture held in the new Moscow Laboratory, Oskar Vogt, the director of the Kaiser Wilhelm Institute, said that deep in the third layer of Lenin's cerebral cortex he found pyramidal cells larger and more numerous than he had ever observed before, and he linked them to Lenin's exceptional powers of associative thought, just as an athlete's strength is linked to his powerfully developed muscles.[34] Modern brain research discounts these conclusions, but at the time they inspired an enthusiastic, popular article in one of the big Berlin dailies by Arthur Koestler, the later novelist, who had embraced the Communist faith.

In exchange for Vogt's services, the Soviet Academy of Sciences promised to help set up in Vogt's institute in Berlin a laboratory of experimental genetics. Vogt owned a large collection of bumble-bees. He was convinced that the different species of bumblebees had arisen by Lamarckian inheritance of acquired characters rather than by mutation and natural selection, and he needed a geneticist to prove his theory. Among the young scientists who were recommended to start work in Berlin was Timoféeff. He left for Germany in 1926 and founded a laboratory for the study of *Drosophila* in Vogt's institute. He never proved Vogt's Lamarckian ideas, but instead became one of the world's leading Mendelian geneticists. Contemporaries describe him as a physical and intellectual giant; in Russia his nickname had been the Wild Boar.

In the 1930s Timoféeff thought of returning to Russia, but his friends advised him that it would be unsafe because Stalin's persecution of Mendelian geneticists had already begun. His younger brothers were arrested, and one was executed. In Germany during the war, his seventeen-year-old son joined an underground anti-Fascist group. He was caught by the Gestapo and disappeared. After the end of the war, when the Russians occupied Berlin, German colleagues advised him to flee to the West, but he decided to stay with his precious collection of flies. In August 1945, Timoféeff was arrested by the Soviet secret police, sentenced to ten years' hard labor, and sent to a prison camp in north Kazakhstan. Later he shared a prison cell with Alexander Solzhenitsyn and twenty-two other prisoners in Bytyrsky. Solzhenitsyn described in *The Gulag Archipelago* how Timoféeff's irrepressible enthusiasm for science made him organize scientific seminars even in that prison cell. Apparently Solzhenitsyn also used him as a model for the scientist in *The First Circle.*

In 1947 the physicist Frédéric Joliot-Curie wrote to L. P. Beria, the head of the Russian secret police, on behalf of the French Academy of Sciences and asked for Timoféeff's release on the grounds that he was a valuable scientist and should be given a chance to do research. Joliot-Curie's intervention saved Timoféeff's life. He had been at death's door, but after several months of recuperation in a Moscow hospital, he was sufficiently restored to set up a new secret prison research institute on radiation biology east of the Urals.

Previously, in September 1945, the Russians had also arrested Karl G. Zimmer with two of his colleagues and taken them to the

Lubljanka Prison in Moscow for interrogation. After some time they were sent to work in a uranium factory not far from the city. When Timoféeff set up his new institute, he asked that Zimmer and his colleagues, as well as his own wife, should be allowed to join him there. Timoféeff's eyesight had been ruined by starvation, and his wife read the scientific literature to him. After Stalin's death they were released from prison but continued to work in Sverdlovsk. In 1964 Timoféeff was asked to organize the Department of Genetics and Radiobiology at the new Institute of Medical Radiology in Obninsk, where Medvedev, the geneticist and author of the famous book *The Rise and Fall of Lysenko* joined him. Medvedev described Timoféeff as a great man and a brilliant scientist; his mastery of many fields of genetics and biology, his dynamism and personal magnetism stimulated the work of the entire laboratory. Peter Kapitsa became his close friend.

Timoféeff retired in 1970 on a pension so meager that it left him almost destitute. He died in 1981, in the same year as his friend Delbrück, who actually came to visit him in Obninsk after having been given the Nobel Prize in Stockholm. But for Schrödinger's book, the name of Timoféeff would have remained unknown outside the circles of genetics and radiation biology.

Zimmer returned to West Germany in 1955. He became one of the first to recognize the importance of electron spin resonance for radiation biology and to prove that ionizing radiations generate free radicals in biological molecules. In 1957 he was offered a chair in Heidelberg combined with the directorship of a new Department of Radiation Biology at the Institute for Nuclear Research in Karlsruhe. There he worked on the effects of ionizing radiations on DNA and other biologically important molecules, and his laboratory became a successful center for fundamental and applied radiobiology. He also published a book on that subject.[35] He died in Karlsruhe in 1988.

As a final irony H. Traut, working in Zimmer's laboratory, found Timoféeff's linear dose-response curve to have been an artifact. He showed that the mutation rate of *Drosophila* germ cells varies widely at different stages of their development. If males are irradiated and then mated, the frequency of mutation among the offspring varies with the time that elapsed between the two events, because the sperm that fertilizes a female five days after irradiation was at an earlier stage of development when it was irradiated than the sperm that fertilizes a female one day after irradiation. At all stages the dose-response curves are nonlinear. Traut demonstrated that a linear response curve, similar to those observed by Timoféeff, is obtained by summing the different dose-response curves produced by matings during the first four days of irradiation.[36]

Zimmer comments, "This result removes one of the foundation-stones of the Green Pamphlet [as the Timoféeff, Zimmer, and Delbrück paper became known]. Strangely enough, that does not seem to matter any more, for two reasons; (i) the concept of the gene and modern trends in genetic research as well as in radiation biology have changed considerably during thirty years, and (ii) the Green Pamphlet has served a useful purpose by helping to initiate these modern trends."[37]

Timoféeff's observation of a fivefold rise in the mutation rate accompanying a 10°C rise in temperature formed the basis for Delbrück's estimate of the energy needed for spontaneous mutations. That observation is now known to have no general validity, because other mutation rates have been found to be independent of temperature or even to drop with rising temperature.[38] These findings remove the other cornerstone of the Green Pamphlet, confirming Karl Popper's dictum that even wrong experimental results may sometimes help the progress of science.

NOTES

EPIGRAPH

1. Jawaharlal Nehru, *Proceedings of the National Institute of Science of India* 27A (196): 564.

PREFACE

1. Peter Medawar, *The Times Literary Supplement* [London], 25 October 1963: 850.

IS SCIENCE NECESSARY?

1. Iris Origo, *The Merchant of Prato* (Harmondsworth: Penguin Books, 1963).
2. H. R. Trevor-Roper, *Religion, the Reformation, and Social Change, and Other Essays* (London: Macmillan, 1967).
3. Martin Gardner, "Seeing Stars," *New York Review of Books*, 30 June 1988.
4. S. C. Brown, *Benjamin Thompson, Count Rumford* (Cambridge, Mass.: MIT Press, 1979).
5. H. L. Gumpert, *Lichtenberg in England* (Wiesbaden: Otto Harrassowitz, 1977).
6. *Seventh Report of the Royal Commission for Environmental Pollu-*

tion, Agriculture, and the Environment (Her Majesty's Stationery Office, Cmd. No. 7644, 1980).

7. M. S. Swaminathan and V. Nagarajan, "Building a National Food Security System," *Indian Journal of Nutrition Science* 16 (Delhi: 1979): 83; M. S. Swaminathan, "Recent Advances in Agricultural Sciences," *Proceedings of a Seminar on Science and Its Impact on Society* (Delhi: Indian National Science Academy, 1978).

8. M. S. Swaminathan, *Global Aspects of Food Production* (Geneva: World Meteorological Organisation, World Climate Conference, 1979).

9. M. S. Swaminathan, "Rice," *Scientific American* 250 (January 1984): 63.

10. Vaclav Smil, "China's Food," *Scientific American* 253 (December 1985): 104; N. R. Lardy, *Agriculture in China* (Cambridge: Cambridge University Press, 1983).

11. *African Agriculture: The Next Twenty-five Years* (Rome: Food and Agriculture Organization, 1986).

12. "A Strategy to Put an End to Starvation," *The Guardian* [London], 9 November 1984.

13. *World Development Report, 1986* (Oxford: Oxford University Press for the World Bank, 1986).

14. M. M. Cernea, J. K. Coulter, and J. F. A. Russell, eds., *Agricultural Extension by Training and Visits: The Asian Experience* (Washington, D.C.: The World Bank, 1983).

15. R. P. Sheldon, "Phosphate Rock," *Scientific American* 246 (June 1982): 31.

16. *Seventh Report of the Royal Commission for Environmental Pollution.*

17. Johanna Döbereiner, J. S. A. Netto, and D. B. Arkoll, "Energy Alternatives from Agriculture," *Pontificiae Academiae Scientiarum Scripta Varia* 46 (1981): 431–58.

18. M. W. Service, "Control of Malaria," in *Ecological Effects of Pesticides*, ed. F. K. Perring and Kenneth Mellanby (New York: Academic Press, 1977).

19. *Seventh Report of the Royal Commission for Environmental Pollution*; Kenneth Mellanby, *The Biology of Pollution*, 2d ed. (London: Edward Arnold, 1980).

20. David Pimentel and Marcia Pimentel, *Food, Energy, and Society*, Resources and Environmental Sciences Series (London: Edward Arnold, 1979).

21. D. C. Wilson, "Lessons from Seveso," *Chemistry in Britain* 18 (1982): 499.

22. D. Weir and M. Schapiro, *Circle of Poison* (Institute for Food and Development Policy, 2588 Mission Street, San Francisco, Calif. 94100: 1982).

23. M. E. Loevinsohn, "Insecticide Use and Increased Mortality in Rural Central Luzon, Philippines," *Lancet* 13 (June 1987): 1359.

24. G. C. Pimental and J. A. Coonrod, *Opportunities in Chemistry* (Washington, D.C: National Academy Press, 1987); R. A. Coffee, "Electrodynamic Crop Spraying," *Outlook on Agriculture* 10 (1981): 350.

25. E. A. Bernays, "Nutritional Ecology of Grass Foliage-chewing Insects," in *Nutritional Ecology of Insects, Mites, and Spiders* (London: Wiley, 1986).

26. *Seventh Report of the Royal Commission for Environmental Pollution.*

27. J. C. Zadoks, "An Integrated Disease and Pest Management Scheme, EPIPRE, for Wheat," in *Better Crops for Food*, Ciba Symposium no. 97 (London: Pitman, 1983).

28. Pimentel and Pimentel, *Food, Energy, and Society.*

29. R. S. Chaleff, *Genetics of Higher Plants* (Cambridge: Cambridge University Press, 1981); *Better Crops for Food*; Michael Bevan, "Binary Agrobacterium Vectors for Plant Transformation," *Nucleic Acids Research* 12 (1984): 8711.

30. Arthur Klausner, "Monsanto: Betting a Giant on Biotechnology," *Biotechnology* 4 (1986): 403.

31. Patricia Powell Abel, R. S. Nelson, Baron De, Nancy Hoffman, S. G. Rogers, R. T. Fraley, and R. N. Beachy, "Delay of Disease Development in Transgenic Plants That Express the Tobacco Mosaic Virus Coat Protein Gene," *Science* 232 (1986): 738.

32. D. M. Shah et al., "Engineering Herbicide Tolerance in Transgenic Plants," *Science* 233 (1986): 478.

33. A. de la Peña, H. Lörz, and J. Schell, "Transgenic Rye Plants Obtained by Injecting DNA into Young Floral Tillers," *Nature* 325 (1987): 274.

34. Max-Planck Institut für Züchtungsforschung, *Max-Planck Gesellschaft, Berichte und Mitteilungen Heft* 2 (München: 1986).

35. M. D. Gale et al., "An Alpha-amylase Gene from *Aegilops ventricosa* Transferred to Bread Wheat Together with a Factor for Eyespot Resistance," *Heredity* 52 (1984): 431.

36. Beatrice Mintz, "Gene Expression in Neoplasia and Differentiation," *Harvey Lectures* 71 (1978): 193.

37. R. D. Pakister, R. L. Brinster et al., "Dramatic Growth of Mice That Develop from Eggs Microinjected with Metallocyanin Growth Hormone Fusion Gene," *Nature* 300 (1982): 611.

38. Edmund Halley, "An Estimate of the Degree of Mortality of Mankind Drawn from Various Tables of the Births and Funerals at the City of Breslau; with an Attempt to Ascertain the Price of Annuities upon Lives," *Philosophical Transactions of the Royal Society* 17 (1693): 596.

39. John Cairns, "The History of Mortality and the Conquest of Cancer," in *Accomplishments in Cancer Research* (Philadelphia: J. B. Lippincott, 1985).

40. Douglas Black, J. N. Morris, C. Smith, and P. Townsend, *Inequalities in Health* (London: Penguin Books, 1982).

41. John Cairns, "The History of Mortality."

42. A. M. Anderson, "The Great Japanese IQ Increase," *Nature* (London) 297 (1982): 181.

43. J. Fry, D. Brooks, and I. McColl, *National Health Service Data Book* (Hingham, Mass.: Kluver Boston, MTP Press, 1987).

44. John Cairns, *Cancer, Science, and Society* (San Francisco: W. H. Freeman, 1978).

45. Fry, Brooks, and McColl, *NHS Data Book*.

46. Richard Doll, Richard Peto, David Evered, and Julie Whelan, eds., *The Value of Preventive Medicine*, CIBA Symposium no. 110 (London: Pitman, 1985).

47. Takashi Sugimura, "Carcinogenicity of Mutagenic Heterocyclic Amines Formed during the Cooking Process," *Mutation Research* 150 (1985): 33.

48. John Cairns, "The Treatment of Diseases and the War against Cancer," *Scientific American* 253 (November 1985): 31–39; Robert W. Miller and Frank W. Mckay, "Decline in U.S. Childhood Cancer," *Journal of the American Medical Association* 251 (1984): 1567.

49. Joan Shenton, "Exporting Danger to the Third World," *The Independent* [London], 23 October 1987.

50. M. F. Steward, "Public Policy and Innovation in the Drug Industry," in *Proceedings of Section 10 (General) of the British Association for the Advancement of Science, 139th Annual Meeting, 1977*, ed. Douglas Black and G. P. Thomas (London: Croom Helm, 1980); H. G. Grabowski, J. M. Vernon, and L. G. Thomas, "Estimating the Effect of Regulation on Innovation: An International Comparative Analysis of the Drug Industry," *Journal of Law and Economics* 21 (1978): 133; *Arzneimittel-forschung in Deutschland* (Pharma, Bundesverband der Pharmazeutischen Industrie, Karlstrasse 21, 6000 Frankfurt: 1979–80).

51. Vulimiri Ramalingaswami, "The People, More Technologies for Rural Health," *Proceedings of the Royal Society* B 209 (1980): 83.

52. Fry, Brooks, and McColl, *NHS Data Book*.

53. Doll, Peto, Evered, and Whelan, *The Value of Preventive Medicine*.

54. D. J. Weatherall, *The New Genetics and Clinical Practice* (Oxford: Oxford University Press, 1985).

55. Bernadette Modell, R. H. T. Ward, and D. V. L. Fairweather, "Effect of Introducing Antenatal Diagnosis on Reproductive Behaviour of Families at Risk for Thalassaemia Major," *British Medical Journal* 1 (1980): 1347.

56. *Report of the World Health Organization/Mediterranean Working Group on Haemoglobinopathies*, Brussels: 14 March 1986; Paris: 20–21 March 1987.

57. H. A. Pearson, D. K. Guillotis, L. Rink, and J. A. Wells, "Patient Distribution in Thalassemia Major: Changes from 1973 to 1985," *Pediatrics* 80 (1987): 53.

58. Thomas Doetschman, R. G. Gregg, Nobuyo Maeda, M. L. Hooper, D. W. Melton, Simon Thompson, and Oliver Smithies, "Targeted Correction of a Mutant HPRT Gene in Mouse Embryonic Stem Cells," *Nature* 330 (1987): 576.

59. Le Roy Walters, "The Ethics of Human Gene Therapy," *Nature* 320 (1986): 225–227; "Points to Consider in the Design and Submission of Human Somatic-Cell Gene Therapy Protocols," *Recombinant DNA Technology Bulletin* 8 (1985): 116–22.

60. R. A. Weinberg, "A Molecular Basis of Cancer," *Scientific American* 249 (November 1983): 102–16; Tony Hunter, "The Proteins of Oncogenes," *Scientific American* 251 (August 1984): 60–69.

61. "The Thrombolysis in Myocardial Infarction Trial: Phase I Findings," *New England Journal of Medicine* 312 (1985): 932–36; M. Verstraete et al., "Randomized Trial of Intravenous Recombinant Tissue-type Plasminogen Activator Versus Intravenous Streptokinase in Acute Hydrocardial Infarction," *Lancet* 1 (1985): 842.

62. A. J. Jeffreys, V. Wilson, and S. L. Thein, "Hypervariable Minisatellite Regions in Human DNA," *Nature* 314 (1985): 67; A. J. Jeffreys, V. Wilson, and S. L. Thein, "Individual-specific Fingerprinting of Human DNA," *Nature* 316 (1985): 76; A. J. Jeffreys, J. F. Y. Brookfield, and R. Semenoff, "Positive Identification of an Immigrant: Test Case Using Human DNA Fingerprints," *Nature* 317 (1985): 818.

63. Lewis Thomas, *The Youngest Science* (New York: The Viking Press, 1983; Oxford: Oxford University Press, 1984).

64. René J. Dubos, *The Professor, The Institute and DNA* (New York: The Rockefeller University Press, 1976).

65. Peter Baxendell, "Enhancing Oil Recovery—Making the Most of What We've Got," *Transactions of the Institute of Mining and Metallurgy* 94A (April 1985): A84–A89.

66. *The Energy Spectrum: Oil, Natural Gas, Coal, Hydro, Nuclear, Biomass, Geothermal, Solar, Tidal, Wind*, Shell Briefing Service, no. 3, 1982.

67. H. W. Lewis, "The Safety of Fission Reactors," *Scientific American* 242 (March 1980): 33; H. M. Agnew, "Gas-cooled Nuclear Power Reactors," *Scientific American* 244 (1981): 43.

68. G. T. Seaborg and J. L. Bloom, "Fast Breeder Reactors," *Scientific American* 233 (1970): 13.

69. *Sixth Report of the Royal Commission for Environmental Pollution, Nuclear Power, and the Environment* (Her Majesty's Stationery Office, Cmd. No. 6618, 1976).

70. Alan Anderson, "Congress Goes for Nevada as Site for Nuclear Waste Storage," *Nature* 330 (1987): 682.

71. N. J. D. Lucas, *Energy in France* (London: Europa Publications, 1980).

72. U.S. Department of Energy, Energy Information Administration, *Electric Power Annual, 1986*. DOE/EIA-0348 (86).

73. U.S. Department of Energy, Energy Information Administration, *Annual Energy Outlook, 1984*. DOE/EIA-0383 (84).

74. "Shutting the Stable Door," *Nature* 223 (1986): 28.

75. "Chronology of a Catastrophe," *Nature* 223 (1986): 28; Richard Wilson, "What Really Went Wrong," *Nature* 223 (1986): 29.

76. William Booth, "Postmortem on Three Mile Island," *Science* 238 (1987): 1342; U.S. Nuclear Regulatory Commission, Office of Government and Public Affairs, Washington, D.C. 20555.

77. *House of Lords Official Report*, 19 November 1986 (Her Majesty's Stationery Office), 348–428.

78. Ibid.

79. Ibid.

80. Walter Marshall, "Tizard Lecture," *Atom*, June 1986, 1–8.

81. A. V. Nero, Jr., "Controlling Indoor Pollution," *Scientific American* 258 (May 1988): 24.

82. David Forman, Paula Cook-Mozaffari, Sarah Derby, Gwyneth Davey, Irene Stratton, Richard Doll, and M. Pike, "Cancer near Nuclear Installations," *Nature* 329 (1987): 499.

83. Guido Biscontin and Luigi Cattalini, "Venice Regained," *Chemistry in Britain* 16 (1980): 360.

84. *Health and Safety Statistics, 1987* (Her Majesty's Stationery Office, ISBBN No. 011883263X, 1981).

85. Roger Revelle, "The Problem with Carbon Dioxide," in *Yearbook of Science and the Future* (Chicago: Encyclopaedia Britannica, 1984).

86. Roger Revelle, "Carbon Dioxide and World Climate," *Scientific American* 247 (August 1982): 33–41; W. S. Moore, "Late Pleistocene Sea Level History," in *Uranium Series Disequilibrium: Application to Environmental Problems*, ed. M. Ivanovich and R. S. Harmon (Oxford: Clarendon Press, 1982).

87. Peter Kapitsa, "Physics and the Energy Problem," *New Scientist* 72 (1976): 10.

88. W. C. Gough and B. J. Eastlund, "The Prospects of Fusion Power," *Scientific American* 224 (February 1971): 50; Gerold Yonas, "Fusion Power with Particle Beams," *Scientific American* 239 (November 1978): 40; Peter Kapitsa, "Energy, the Fusion Solution," *New Scientist* 72 (1976): 83.

89. Martin Ryle, "The Energy Problem," *Resurgence* no. 80 (May–June 1980): 6; M. Spencer, "Nuclear Energy, the Real Cost," *Ecologist* [London] (1982).

90. D. W. Davidson, "Methane Hydrates," in *Natural Gas Hydrates*, ed. J. L. Cox (Boston: Butterworth, 1983).

91. Thomas Gold, *Power from the Earth* (London: J. M. Dent, 1987).
92. *World Development Report, 1986.*
93. L. R. Brown, "World Population Growth, Soil Erosion, and Food Security," *Science* 214 (1981): 995.
94. Carl Djerassi, *The Politics of Contraception* (New York: W. W. Norton, 1979).
95. Karl Popper, *The Open Society and Its Enemies* (London: Routledge & Kegan Paul, 1962).
96. Peter Medawar, "Induction and Intuition in Scientific Thought," in *American Philosophical Society Memoirs* 75, Jayne Lectures (Philadelphia: 1969).
97. Götz Aly, ed., *Aktion T4 1939–45: Die "Euthanasie"-Zentrale in der Tiergartenstrasse 4* (Berlin: Edition Hentrich, 1987). My translation.
98. Ibid.; Benno Müller-Hill, *Tötliche Wissenschaft: die Aussonderung von Juden, Zigeunern und Geisteskranken, 1933–1945* [Deadly Science: The Selection of Jews, Gypsies, and Mental Patients, 1933–1945] (Rowolt Taschenbuch Verlag, Postfach 1349, D-2057 Reinbeln bei Hamburg, 1984).
99. Müller-Hill, *Tötliche Wissenschaft.*
100. S. A. Fetter and K. Tsipis, "Catastrophic Releases of Radioactivity," *Scientific American* 244 (1981): 33.
101. Solly Zuckerman, *Nuclear Illusion and Reality* (London: Collins, 1982).

ENEMY ALIEN

1. Andrew Boyle, *The Climate of Treason* (London: Hutchinson, 1979).
2. Peter Gillman and Leni Gillman, *Collar the Lot!* (London: Quartet Books, 1980).
3. Uberto Limentani, "Survival at Sea" in *Magdalene College Magazine & Record* (Cambridge, 1980–81): 41–48.

ATOM SPY

1. David Holloway, *The Soviet Union and the Arms Race* (New Haven: Yale University Press, 1983).

DISCOVERERS OF PENICILLIN

1. Gwyn Macfarlane, *Howard Florey: The Making of a Great Scientist* (Oxford: Oxford University Press, 1979).

2. Ronald Hare, *The Birth of Penicillin* (London: George Allen & Unwin, 1970).
3. Ibid.
4. Ibid.
5. R. W. Clark, *The Life of Ernst Chain* (London: Weidenfeld & Nicolson, 1985).
6. Sir Edward Abraham, "Ernst Boris Chain," in *Biographical Memoirs of Fellows of the Royal Society* 29 (1983): 43.
7. André Maurois, *The Life of Alexander Fleming* (London: Jonathan Cape, 1959).
8. Peter Medawar, "Induction and Intuition in Scientific Thought," in *American Philosophical Society Memoirs* 75 (1969).
9. Sanford Brown and Benjamin Thompson, *Count Rumford* (Cambridge, Mass.: MIT Press, 1979).

CHEMIST INTO STATESMAN

1. Chaim Weizmann, *Trial and Error* (New York: Harper Brothers, 1949).
2. Isaiah Berlin, *Chaim Weizmann* (London: Weidenfeld & Nicolson, 1958).
3. Barnet Litvinoff, ed., *The Essential Chaim Weizmann* (London: Weidenfeld & Nicolson, 1982).
4. Ibid.
5. Ibid.
6. Ibid.
7. Berlin, *Chaim Weizmann*.
8. Litvinoff, *The Essential Chaim Weizmann*.
9. Berlin, *Chaim Weizmann*.
10. Litvinoff, *The Essential Chaim Weizmann*.

HOW TO BECOME A SCIENTIST

1. Baruch Blumberg, *Les Prix Nobel* (Stockholm: 1976).

BRAVE NEW WORLD

1. R. C. Haddon and A. A. Lamola, "The Molecular Electronic Device and the Biochip Computer: Present Status," *Proceedings of the National Academy of Sciences* 82 (1985): 1774–1878.
2. Freeman Dyson, *Disturbing the Universe* (New York: Harper & Row, 1979).
3. Steven Weinberg, *Discovery of Subatomic Particles* (New York: Scientific American Books, 1983).

NATURE'S TINKERING

1. August Weismann, *Essais sur L'Hérédité* (Paris: C. Reinwald et Cie, 1892).

DARWIN, POPPER, AND EVOLUTION

1. Karl Popper, *Conjectures and Refutations* (London: Routledge & Kegan Paul, 1972).
2. Karl Popper, *The Open Society and Its Enemies* (London: Routledge & Kegan Paul, 1962).
3. Christian Bauer, H. S. Rollema, H. W. Till, and Gerhard Braunitzer, *Journal of Comparative Physiology* 136 (1980): 67.
4. D. Petschow, Irene Würdinger, Rosemarie Baumann, G. Duhm, Gerhard Braunitzer, and Christian Bauer, *Journal of Applied Physiology* 42 (1977): 139.
5. M. A. Chappell and L. R. J. Snyder, *Proceedings of the National Academy of Sciences* 81 (1984): 5484.
6. Jonathan Flint et al., *Nature* 321 (1986): 744.
7. Josée Pagnier et al., *Proceedings of the National Academy of Sciences* 81 (1984): 1771.

DEPARTMENT OF DEFENSE

1. C. Kumar Patel and Nicolaas Bloembergen, cochairmen, "Report of the American Physical Society on the Feasibility of Directed Energy Weapons," *Scientific American* (June 1987):16.
2. F. Noireau, "HIV Transmission from Monkey to Man," in *The Lancet* (27 June 1987): 1499.
3. A. Kashamura, *Famille, Sexualité et Culture* (Paris: Payot, 1973).

MORE ABOUT IMMUNITY

1. Peter Medawar, *The Times Literary Supplement* [London], 25 October 1963: 850.

PHYSICS AND THE RIDDLE OF LIFE

1. N. W. Timoféeff-Ressovsky, K. G. Zimmer, and Max Delbrück, *Nachrichten aus der Biologie der Gesellschaft der Wissenschaften Göttingen* 1 (1935): 189–245.
2. E. J. Yoxen, *History of Science* 17 (1979): 17–52.
3. François Jacob, *The Logic of Living Systems* (London: Allen Lane, 1974).

4. F. H. C. Crick, *British Medical Bulletin* 21 (1965): 183–86.

5. F. H. C. Crick, quoted by R. C. Olby, *Journal of the History of Biology* 4 (1971): 119–48.

6. E. J. Yoxen, *History of Science* 17.

7. G. W. Beadle and E. L. Tatum, *Proceedings of the National Academy of Sciences* 27 (1941): 499–506.

8. J. B. S. Haldane, *The Biochemistry of the Individual in Perspectives of Biochemistry*, ed. J. Needham and D. E. Green (Cambridge: Cambridge University Press, 1937), 1–10.

9. C. H. Waddington, *Nature* 221 (1969): 318–21.

10. Joseph Weiss, *Nature* 153 (1944): 748–50.

11. E. Collinson, F. S. Dainton, D. R. Smith, and S. Tazuke, *Proceedings of the Chemical Society* (1962): 140–44.

12. G. Czapski and H. A. Schwartz, *Journal of Physical Chemistry* 66 (1962): 471–79.

13. F. S. Dainton, *Chemical Society Reviews* 4 (1975): 323–62.

14. Niels Bohr, *Nature* 131 (1933): 458–60.

15. Linus Pauling and Max Delbrück, *Science* 92 (1940): 77–79.

16. Haldane, *The Biochemistry of the Individual.*

17. Ludwig Boltzmann, *Der zweite Hauptsatz der mechanischen Wärmetheorie* (Vienna: Sitzungsber ichte der Kaiserlichen Akademic der Wissenschaften, 1886).

18. E. J. Yoxen, *History of Science* 17.

19. Ibid.

20. O. T. Avery, C. M. McLeod, and Maclyn McCarty, *Journal of Experimental Medicine* 79 (1944): 137–58.

21. J. B. S. Haldane, *Nature* 155 (1945): 375–76.

22. Arthur Kornberg, *DNA Replication* (San Francisco: W. H. Freeman, 1980); and Kornberg, *Supplement to DNA Replication* (San Francisco: W. H. Freeman, 1982).

23. Olga Kennard, "Structural Studies of Base Pair Mismatches," in *Structure and Expression, DNA and Drug Complexes*, eds. R. H. Sarma and M. H. Sarma (New York: Adenine Press, 1988):1–25.

24. Kornberg, *DNA Replication*; Kornberg, *Supplement.*

25. A. R. Fersht, *Proceedings of the Royal Society* B 212 (1981): 351–79.

26. B. W. Glickman and Miroslav Radman, *Proceedings of the National Academy of Sciences* 77 (1980): 1063–67. For a review of mismatch repair in *E. coli* see Miroslav Radman and Robert Wagner, *Annual Review of Genetics* 20 (1986): 523–38; *Scientific American* (August 1988): 24.

27. Eduard Batschelet, Esteban Domingo, and Charles Weissman, *Gene* 1 (1976): 27–33.

28. Linus Pauling, *Festschrift Prof. Dr. Arthur Stoll Siebzigsten Geburtstag* (1958), 597–602.

29. Fersht, *Proceedings of the Royal Society.*

30. R. B. Loftfield and Dorothy Vanderjagt, *Biochemical Journal* 128 (1972): 1353–56.
31. Fersht, *Proceedings of the Royal Society.*
32. J. J. Hopfield, *Proceedings of the National Academy of Sciences* 77 (1974): 4135–39; R. C. Thompson and P. J. Stone, *Proceedings of the National Academy of Sciences* 74 (1977): 198–202; J. L. Yates, *Journal of Biological Chemistry* 254 (1979): 1150–54.
33. Z. A. Medvedev, *Genetics* 100 (1982): 1–5.
34. Oskar Vogt, *Journal für Psychologie und Neurologie* 40 (1929): 108.
35. K. G. Zimmer, *Quantitative Radiation Biology* (Edinburgh: Oliver & Boyd, 1961).
36. H. Traut, "Dose-Dependence of the Frequency of Radiation-induced Recessive Sex-linked Lethals in *Drosophila melanogaster*, with Special Consideration of the Stage Sensitivity of the Irradiated Germ Cells," in *Repair from Genetic Radiation Damage*, ed. F. H. Sobels (London: Pergamon Press, 1963), 359.
37. K. G. Zimmer, "The Target Theory," in *Phage and the Origins of Molecular Biology*, eds. John Cairns, G. S. Steng, and J. D. Watson (Long Island: Cold Spring Harbor Laboratory of Quantitative Biology, 1966), 33–42.
38. B. L. Sheldon and J. S. F. Barker, "The Effect of Temperature on Mutation in *Drosophila melanogaster*," *Mutation Research* 1 (1964): 310–17.

INDEX

In this index, the letter *f* or *t* following a page number indicates a reference to a figure or a table, respectively.